"To those who know him, Norm Hackerman is famous because he seems to know everything, and because he can do several jobs simultaneously better than most of us can do one job. But Hackerman is especially known for his straight talk and "telling it like it is." *Conversations* is quintessential Hackerman–in a short book he and Ashworth take the reader from the beginning of science to modern times, all the while cutting through the fog and baloney about science and technology that often make it seem beyond the reach of the average person. This enjoyable book should be required reading for those politicians and others who would question the value of science and technology to modern society and the future of our planet."
—**Richard S. Nicholson, Executive Officer, American Association for the Advancement of Science**

"An issue of great importance to the country has been steadily emerging concerning the strength of the federal support for science and technology, particularly in basic and applied research. *Conversations on the Uses of Science and Technology* goes right to the heart of the issue in addressing the uses of science and technology. And what an interesting and informative approach it is–interesting because of the easy-to-read, platonic dialogue used by the authors, informative because of the authors' combined experience and knowledge of scientific and technologic activities in universities, industries, and government, at both the professional practice and the policy levels, for example, Norman Hackerman's two term Chairmanship of the National Science Board. The central issue is attractively enhanced by a description in lay terms of the nature of science and technology, and the characteristic activities of research and development.

Conversations is recommended reading for anyone and everyone with an interest in the health of science and technology—students, professors, researchers, administrators in the universities, members of Congress and leaders of the Administration, as well as interested members of the voting public."

—**H. Guyford Stever, Former Director,
National Science Foundation**

"This wonderful and timely book is pure Norman Hackerman. Dr. Hackerman has as many credentials to talk about the importance of science and science policy as any person alive today.

He makes the case for basic research and education without falling into the usual arguments and clichés. These gentlemen will educate anyone, whether technically competent or not. This book will influence politicians and policy makers in governments and universities."

—**Herbert D. Doan, Retired President,
The Dow Chemical Company**

CONVERSATIONS
ON THE USES OF
SCIENCE AND TECHNOLOGY

CONVERSATIONS
ON THE USES OF
SCIENCE AND
TECHNOLOGY

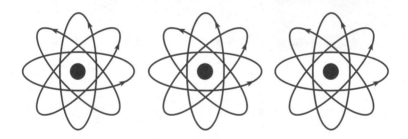

Norman Hackerman and
Kenneth Ashworth

University of North Texas Press
Denton, Texas

© 1996, Norman Hackerman and Kenneth Ashworth
Manufactured in the United States of America
All rights reserved

10 9 8 7 6 5 4 3 2 1

Requests for permission to reproduce any material from this work should be sent to:

Permissions
University of North Texas Press
P.O. Box 13856
Denton TX 76203

The paper in this book meets the minimum requirements of the American National Standard for Permanence of Paper for Printed Library Materials, Z39.48.1984

Library of Congress Cataloging-in-Publication Data

Hackerman, Norman.
Conversations on the uses of science and technology / by Norman Hackerman & Kenneth Ashworth.
    p. cm.
  ISBN 1-57441-015-6
    1. Scientists—Interviews. 2. Science—Social aspects. 3. Technology—Social aspects. 4. Ashworth, Kenneth H.
I. Ashworth, Kenneth H. II. Title
Q141.H2164 1996
500–dc20   96-19162
                                                                                 CIP

Design by Amy Layton

Table of Contents

Foreword	vii
Preface	ix
Acknowledgments	xi
Early Science and Technology	1
Training versus Education	13
Science and Technology Today	29
Technology Transfer	55
Science and the Non-scientist	71
Big Science and Little Science	85
Index	105

Foreword

If you like listening to an interesting and exciting conversation between two knowledgeable and literate individuals, you will enjoy the dialogue between Norman Hackerman and Kenneth Ashworth. The conversation has all of the hesitancy, groping for words and phrases and concepts, that typifies casual conversation. The dialogue is a very personal exposition of Norman Hackerman's views of science, technology and its interaction with education and general public policy. In his inimitable way, Dr. Hackerman communicates the great wisdom about such matters for which he is well known. Some of his personal views on science, technology and society are at variance with some of the more widely expressed mainstream views.

Norman Hackerman should know whereof he speaks, because for many years he has been deeply involved in the issues which he discusses. As Chairman of the National Science Board (which governs the National Science Foundation) from 1974 to 1980, he observed and participated in the formulation of science policy. He brings to the conversation the insight of an individual who has been there. As the past president of Rice University, he expresses the highest principles and values about the key role of education. He makes sense for the reader of much that seems odd and mysterious about science and technology. What we have in this dialogue is vintage Norman Hackerman; views that are always interesting and challenging, and insights into the thinking of an American original.

Robert M. White
President
National Academy of Engineering

Preface

Having heard Norman Hackerman testify before legislative committees and other groups as he espoused his views on the dependency of mankind upon science and technology, I asked him why he had never put his ideas in print. He begged off as too busy. As I continued to press him over time, he then claimed to be unable to read his own notes. (A statement totally certifiable by any of Norm's friends or former secretaries.)

To show him he already had a good beginning for an article, I had two of his presentations transcribed. Norm looked them over and agreed to talk with me over lunch about the issues he had raised. Thus, in January 1995, began a series of luncheon meetings with a small recorder lying in the center of our table. The topics we discussed grew over the months we met and talked. From out of the recorded clatter of dishes, interruptions by waitresses, and the background babble, I extracted the conversations presented here.

Friends have asked, "Where did you get your questions? How did you know where you wanted to go in your discussions?" Those were never my concern. My worry at every luncheon was whether the recorder was working. What made the discussions most informative and engrossing was their spontaneity; everyone knows Norm does not like to plow the same ground twice. I knew I would best get the real Norman Hackerman on the first take.

As to where we were headed—we followed our noses, and talked about what we had been reading or how good public policy was being damaged or threatened.

When I would apologize for my inability to follow one or the other of Norm's arcane scientific explanations he would wave that off and say that, quite the contrary, it was perfect for what we were trying to do. If I could not understand him, neither could the principle audience we hoped to reach. Then he would take another approach to help me understand.

So in the end we rambled a lot. But in the process we talked about our place in the universe and in time, the uniqueness of *Homo sapiens*, how we got to where we are in our history, how so many of us in recent years have managed to stay alive and to exist as well as we do, and the necessary conditions for us to continue to do better—at least for a little while, if we understand and apply how the rules of nature work. A key question was whether the recent "golden age" of science can be extended. What could be more intriguing—for a scientist or a nonscientist?

Kenneth Ashworth

Acknowledgments

For early editorial assistance on three chapters we are grateful to Michele Kay. For assistance in tracking down sources and documentation and for general encouragement we are indebted to Malcolm Gillis, William Livingston, Gerhardt Fonken, Paul McClure, and Sidney Stewart. For suggestions for additional topics we thank Peter Flawn. For research in the libraries at the University of Texas we are grateful to Dorothy Carner, Nancy Green, and Lorraine Dallas. For preparation and corrections we are grateful to Shayne Hansen and Mary Allen. Thanks also to Charlotte Wright and Frances Vick at the University of North Texas Press.

"Nature to be commanded must be obeyed."
—Francis Bacon

# Early Science and Technology

KA: We often think casually of science as having always been here, just as we now unthinkingly assume that radio and recordings and soap and chairs accompanied mankind as we evolved. We typically think of science as one of the accouterments that came along with us, like the clothes it seems we have worn from the very beginning. Maybe we could start with this.

NH: It doesn't matter where we start. We'll end up at the same place. In the beginning there was only nature, precisely the same nature that exists in the universe today. No different. There was just no science. Our earliest ancestors didn't know much about how nature worked. That's why they spent so much time trying to placate the gods of volcanoes and the weather. An eclipse of the sun or moon obviously seemed like a message from the gods.

The concept of science, as knowledge about nature, began modestly as some early men and women began to develop and recognize uses for items from nature.

Once an item became useful, somebody began to ask: What is that thing? Where did it come from? Where can we find some more of it? How can we make another one? Is there some way to make it better or to do it with less effort?

Over time, people worked their way backwards from questions about using things in nature to beginning an accumulation of knowledge about nature. It's at that point that we can say they began to create science—which is nothing more than a systematic understanding of nature. Humanity at that point began working at the mouth of the stream of the flow of knowledge and information.

In the beginning, the flow in that stream was a mere trickle. Humanity, over time extracting uses from that stream, has exerted pressures for more knowledge until today that stream has become an enormous flowing river. And those individuals who have pushed their way up that river to increase the flow of knowledge at the headwaters have become our scientists.

KA: We're moving fast. You've just taken us tens if not hundreds of thousands of years in just a few sentences.

NH: Maybe so, but even today most of mankind still functions at the mouth of that river of knowledge, drawing on knowledge for the purpose of finding the *uses* to which it can be put.

The very word "science," in its Latin source, *scientia*, means knowledge. Today a few individuals function to a large extent upstream, serving at the headwaters to generate knowledge and understanding about nature and the universe. The expectations downstream for a faster flow of information and knowledge have continued to grow unabated. In fact, the expectations and demand have greatly increased. Economies, political regimes, ideologies, entire civilizations and, today, almost the entire planet, have all become dependent on the knowledge flowing from science and how that knowledge is translated and transformed into products and jobs and uses by people all over the world.

KA: Let's go back and speculate for a bit. What do you think might have been some of the first primitive uses of science?

NH: I don't know. Remember, we're talking about knowledge about nature. I'd guess maybe breeding. Take shepherds, for instance. One of them somewhere in all those years of watching flocks might've looked at two especially good goats or sheep and then made the leap to ask the right question.

Or let's say somebody noticed that an outsider with dark hair and dark eyes who stayed within a blond tribe ended up with dark haired and dark eyed children. Can't you imagine all the people gathering around the babies back then just the way they do today and saying, "Oh, look how he favors his father—he's got his eyes, his coloring," etc. You know how it goes. And then up in the hills one day some idle shepherd might say, "Hey, I wonder what would happen if I put those two particular goats together." Or those two dogs.

As a matter of fact, when you think about it, it might've been with dogs first. The difference in crossing dog breeds shows up so fast. And being able to see particular parental traits of different dogs in the same litter would have to impress itself even on people with small brains—but I'm only kidding. By the time men were shepherding animals, mankind was already as fully brained as we are today.

KA: And you'd call that science, primitive animal breeding?

NH: Wouldn't you? It's a piece of knowledge out of the swamp of ignorance. It's something you can repeat and get the same results a good bit of the time.

KA: But let's take something that we think of more commonly today as science. Let's say some primitive use of materials sciences. How about finding and using natural or native materials?

NH: There aren't many native materials. Obsidian or flint were worked by chipping and flaking, and archaeologists find sites where the tailings show men squatted and shaped their tools.

KA: How about metal materials?

NH: There aren't many that occur naturally. Gold does. And there's native silver. There's very little native copper, but some. Not much else. Pure iron shows up only in meteorites. The reason is the elements are too active chemically. The more active an element is, the less likely it is to show up in native or pure form geologically. They are combined with other elements and they take form in nature as compounds.

KA: So man somewhere along the way had to find out how to separate the elements he wanted from other elements they were combined with in nature.

NH: Exactly. The compounds in nature are almost always mixed with other naturally occurring compounds. Those are mixture ores. They are typically of one or more metallic elements. And early humans began to find uses and eventually realized there could be economic gains from extracting some of those metals. With some few elements they could do that by heat. That's the way early man got iron out of the ore they found it in and gave all of us the next leg up from the Bronze Age, which had already been going on in a few locations for over 2500 years. Most of us wouldn't be here if somebody very early hadn't figured out how to do that with iron.

KA: When my wife and I were in Bodrum, on the west coast of Turkey, we saw some of the materials brought up from under the Mediterranean from ancient ship wrecks. Among the most peculiar findings were dozens of heavy casts of some kind of ore, flat and shaped sort of like huge H's, maybe thirty inches high, with the bulk of the ore in the crossbar. When we asked what they were, we were told they were copper. When we asked about the peculiar shape, the answer came back: "How would you design something that weighed around eighty to a hundred pounds that you'd processed in the mountains and that you had to strap on a mule and haul over long, rough trails to a seaport? It wouldn't be round and it wouldn't be square or a cube or a compact ingot. Think about it; it would be shaped like an H to give you some handles. No other design would work better."

So you're saying that copper like that we saw, which was part of the key to the Bronze Age, had to have been processed and separated from a more complex ore?

NH: Right. And maybe they did that painstakingly by chipping and picking at the ore and applying heat to the richest pieces they had left. Or they could have done it by soaking that raw ore in some kind of acid—sulphuric or hydrochloric or nitric. Now how acids got discovered or by whom I haven't the first idea. But we do know they had to use heat or acids. That much is inescapable. Somehow, somebody at some point discovered that by using heat or an acid solution they could get materials they wanted to separate from the debris in a composite ore.

KA: Any speculations on where they got the acids?

NH: Well, in more modern times, Cortez on his way to Tenochtitlan, the Aztec capital, sent his soldiers down into the smoking crater of Popocatépetl to get sulphur. But looking in volcanoes for certain elements was old hat by his time. In volcanoes you can find elements that are geologically young, just being formed out of the earth's magma. There I'm sure you can find SO_2 and when you dissolve that in water you get sulphurous acid, and if you let it sit around long enough you'll get sulphuric acid. Or if you soak iron in neutral sodium chloride, a byproduct is hydrochloric acid . . .

KA: And where would primitives have found sodium chloride?

NH: Salt beds or salt extrusions, perhaps. Salt licks have always attracted animals and people. And then you might find nitrous or nitric acid naturally after a thunderstorm, since nitric oxide forms as a consequence of lightning. On further thought this would probably not create enough nitric acid in nature to be used for separating elements of the magnitude we're talking about.

But even if I don't know how they found out about using acids to separate elements, it is clear that somebody, sometime discovered the effects of acids on elements they found in the wild. I don't know how or when, but that's not important to my concern, which is the transfer of knowledge to technology in order to improve the welfare of mankind.

I do know you don't find native lead, or nickel, or zinc lying around in any significant quantity. And if you want to make bronze and make a big leap for mankind from the stone age, you must mix copper and tin. There's no escaping that. Knowledge of nature had to be applied to the technology that first brought about the

Bronze Age, and then a couple of thousand years later the Iron Age.

It's clear that out of the ignorance surrounding all that inert ore and those early human beings, some individual or team of individuals had to discover enough about how nature works so they could extract copper from its other natural detritus. Then I guess after some of them got pretty good at it, they melted what they extracted and poured it into sand molds, to shape it, as you said, in ingots shaped like H's, for transport to market to barter or make a profit.

Just how they found all that out I can't tell you. But we have to remember, they had lots of time. It's been only about 6,000 years since some primitive scientist set us on the Bronze-Age road from where we were in the Stone Age. And our forebears before them had hundreds of thousands of years to tinker with ignorance before hitting on their important discovery that started the use of stone tools, first chipping and then grinding.

In any event, the scenario I've just made up has all the elements of what had to come together in some form, or civilization on this planet would still be stuck in the Stone Age. And for the race to have survived a couple of ice ages lasting tens of thousands of years with practically no technology must have made it touch and go at times as to whether we were going to be here today.

Incidentally, you only asked about metals materials. An equally important materials technology was early ceramics. Think what it meant to primitive people to discover how to fashion pots and containers to use for transporting and storing water, or protecting their foods from rodents and insects. This had to be one of the race's greatest advancements, and it is probable that women discovered it. And how did they learn to "fire" their pottery? Probably by accident, just as we still learn new things about nature. In any event it was new and valuable knowledge, about how water and certain soils would stick together, what they did when they dried, and what fire did to them.

But of this much I can assure you: without science and technology the world population would be a mere fraction of what it is today. For it is certain that this planet cannot support six billion people without the continuous contributions of science and technology.

KA: Now we truly have moved hundreds of thousands of years in a few moments. From those primitive early contributions by science we are today at the stage where many people believe that even mankind's far more complex and sophisticated needs and wants can be fulfilled by science upon demand. Some people believe that the needs of society can now dictate to science what knowledge and discoveries should be pursued.

NH: That kind of "science on demand" is what impatient managers and executives and politicians hope to get when they invest in research. But that kind of thinking about science is based on a totally false premise, and it is important to spend a minute to understand why it is wrong. We need to understand why investing in directed research or in "instructed" science focused on problems and expected outcomes will be a failure if carried too far. This approach will not only leave the managers and politicians disappointed, but it can actually do harm to science.

KA: But it seems so natural to these people with the money and the authority to give orders saying: "Well, here's the problem; get the facts, find a solution, and we'll put some money into implementing it. What's so complicated about that?"

NH: That is linear thinking. Nature isn't organized to reveal itself to us on demand, or in a linear mode, or in pieces readily or obviously related to one another. And the laws of nature are cold and indifferent to the arrogance of mankind defining, at this moment in the universe, what he needs from nature to most benefit himself on this planet. Accurate knowledge about nature is, for the most part, turned up randomly—stochastically, to use a fancy word for it.

KA: So what should we be after?

NH: I'd say it's this: protecting and advancing civilization and a humane existence on this planet in the face of runaway population growth. That would be a good start. Because without advancing technology, there is no conceivable way we can sustain our growing populations. Without good science, technology will slowly grind to a halt. Without productive economies, we won't have stable governments or peaceful tribes of people. We'll have total chaos since we'll all be fighting for the little that's available for far too many people. So like it or not, we are heavily dependent on well educated people and good science.

KA: And that obviously involves looking at policy areas that connect economic development to knowledge, science, and technology. Businessmen want new products, manufacturers want new materials and processes. Politicians want more jobs for their constituents. And they all look to science and technology for their answers.

NH: That's why Daniel Boorstin's phrase "the illusion of knowledge" is so fitting to what we're talking about. Everyone thinks they understand the role of science in our advancement through the ages because they look upon the blessings and outpourings of science through the benefit of hindsight. Yet when it is in the beginning, experimental stages, every advance made by mankind is an unknown. It is all a gamble, a game whose outcome is at least partly unpredictable. That's why science can be *directed* only in small part. The pieces of knowledge about nature and its uses only *seem* to fit together neatly as we look back on the event of a discovery or an application to a human use.

KA: In that sense science is the same as biography. Every life story is only a series of events whose outcomes were unknown as they were lived. Only by hindsight do they seem to make up a predestined life. Well, clearly we need to reexamine some of our basic assumptions—especially those that are on the verge of being used to set public policy for investments in scientific research. There must be some way to get policy makers to become at least somewhat uncertain about some of the things they seem most certain about.

NH: Perhaps we ought to use a subtitle for our conversations: "Muddling Around with Science." If we can get a few more people to understand science and its uses, maybe our discussions will serve a useful purpose. If we are successful in promoting a better understanding of the issues and processes of science, maybe there would be fewer outsiders muddling around *with* science so that those *inside* science can get on with the muddling around that is necessary in our search for new knowledge.

I am referring, of course, to the scientists' need to follow their noses—where their investigations take them—not where someone directs them to go. It's not a neat process, and I want to come back to that later.

# Training Versus Education

KA: Let's face it, the principal reason state legislators and congressmen support science, research, technology, and development is to improve the economic viability of individual states or perhaps of the country as a whole.

NH: Of course. Many of them are not primarily interested in supporting higher education for some intrinsic worth it may have or as the repository of knowledge or as a cultural institution. The primary interest of most people involved in the political process as it pertains to higher education is its importance as a tool in the economic future of his or her state or as a way to prepare young people for careers and jobs. Legislators may actually recognize that economic viability and vitality alone can't create a happy society, but they clearly do believe that economic strength is a necessary *condition* for maintaining a happy and stable society.

KA: But it goes beyond our own country. Worldwide, policy makers everywhere recognize that science, technology, research, and development are essential to the welfare of mankind—and their constituents.

NH: Look, let's be even more blunt about this. A world which is about to reach six billion people, and with the number growing daily by millions or tens of millions, cannot exist without technology. There is not a prayer of a chance of keeping over six billion people alive on this planet, much less moderately happy or peaceful, without technology.

KA: Clearly, without technology nature would seek a new balance, revert humankind to a much reduced population.

NH: Absolutely. And nature has never been kind in how it does that. As some poet put it, "Nature, red in tooth and claw." We cannot wrest from the planet all the things we need in terms of food, shelter, health, energy, and other types of support for that many people in the old inefficient ways our ancestors did.

As a result, in a very real sense we do not have any choice but to constantly maintain and advance technology to keep up with our almost exponential population increases.

KA: Or our increasing wants and needs—which in many places outpaces our growth in numbers.

NH: Good point; worldwide, material expectations are rising. But aside from providing basic support for abundant human life on this planet, technology is important, at the level of national policy, for two other reasons as well. The first I have already mentioned—its importance to economic competitiveness. The other, of equal importance to political leaders, is the security of the nation. Probably the most spectacular example of that was the American development of the nuclear bomb in World War II, largely out of a fear that Nazi Germany might develop it first.

KA: All right then, what are the prerequisites to enable technology to advance continuously?

NH: Well, first, technology is not merely the presence of resources. The blessings of good climate, fertile land, or abundant mineral deposits are not technology. Yet, if a localized region of the planet, such as a nation or a state, is to become economically viable, it has to have at least one very important natural resource. It is always highly advantageous to have more than one—as is the case with the United States—but it is essential that a region or nation have at least one major resource to acquire any economic significance. Historically, there have been regional economies that have flourished or at least been economically viable based on agriculture alone. In fact, from the earliest civilizations until this century, farmers made up by far the largest group of workers, and in less advanced societies that is still the case.

Agriculture has the advantage of being dependent upon a renewable resource, but it is not a resource subject to high commercial leveraging because many other regions also can provide their own agricultural support or can compete with agricultural products. In addition, agriculture tends, over time, to drift toward non-renewability as land wears out and water resources become limited, especially if agriculture is carried on with limited applications of technology.

Historically, mineral resources have also supported localized regions throughout human history. The state of Texas, for example, enjoyed for over seventy years the benefits of extracting and selling the minerals which fate had endowed to the state, and these were extracted and used at a time when they were valuable to the state's progress.

KA: Can you be more clear on that last comment?

NH: For thousands of years, beneath what we now call "Texas," oil and gas and sulphur lay undetected by the humans living on the land. Only recently have people developed both the methods for extracting them and the technological uses for them.

Today, countries in the Middle East are in a similar situation. There are now and will continue to be many people in those countries and sheikdoms who live well on extracting and exporting oil and gas. Similarly, the extraction of coal, iron ore, or copper has supported other localized areas on the planet. The problem with mineral resources is that they become depleted over time and are not renewable within contemporary times or necessarily in the same location. And even to the extent that minerals like iron and copper are reusable, the reuse nearly always benefits some region other than the one where the minerals originally were extracted. That doesn't have to be the case, but it usually is.

KA: That doesn't present a very promising future. The resources upon which ranching, agriculture, and mining depend all deplete to some extent, perhaps entirely. What's left?

NH: I'm coming to that. There is, in fact, only one resource which is totally renewable and which, in the context of the needs of technology, is a major leveraging component. That is the human resource. It is, in fact, the human resource which should be the major focus of political leaders of any nation or localized region which hopes to advance its economic base. While the important subjects appear to be science, research, technology, and development, none of these has any meaning without focusing on the development of the *human* resource which underpins all of these. How we develop this human resource so as to provide the capabilities and technology for our nation—or any state or region—in order to stimulate economic viability and competitiveness, is the fundamental issue facing our leaders, whether they recognize that or not.

KA: So obviously you bring us to a justification for investment in education and training.

NH: You're pushing me again. I'm not going to say that the development of human resources to maintain technological advancement is a totally adequate definition or justification for education. And you just distinguished between education and training. That's an important distinction that I'll come back to in a minute.

I am talking now about how technology supports the economic advancement of a nation or region, about how technology is dependent on science, and about how both of them are dependent on competent and productive researchers and operational scientists and engineers. The happiness of the society and the economic surplus required to maintain the society are fundamentally dependent upon the economic viability resulting from technology and those supporting that technology. Therefore, I believe a special case can be made for the development of the *individual* over all other resources as the principal component in the advancement of technology.

When I argue that the human resource is renewable and must be developed as the principal leveraging component of technology, let me begin by reminding you of what happens when humans are left to fend for themselves without the benefits of technology. That is, let's consider the situation of human beings living as they have done throughout most of their existence on this planet—taking care of themselves in a fairly simple and primitive way within whatever setting they happen to be born into. We still see examples of this situation in a number of countries, even those which have "developed," but not yet to the point of providing education and training for all their people. Without training or education—without development of their unique human resources—people in any society are left largely to fend for themselves in their raw environment, attempting to eke out some marginal existence from the raw materials on or near the surface where they happen to live. For eons, that was how our forebears survived, using the technology of stones and then a few minerals to form bronze or iron tools. And, as I said before, those advances were very recent in man's long existence on this planet.

The differences made by modern technology in advanced countries like Great Britain, Japan, Singapore, Taiwan, Sweden, and South Korea are self-evident. I pick those particular countries because they have relatively limited natural resources, and yet the development of human resources and potentialities has made technology work to the enhancement of their economies and the quality of life. There are not many resources lying around on the surface in those countries that would long support their standard of living or their present populations without modern technology.

After considering how our primitive forebears were able to extract only a marginal existence from the earth even when they had abundant natural resources all around them, we can today observe how those countries with practically no natural resources whatsoever have become extremely successful economically. They have made that leap almost exclusively by education and training of human resources. This has made a singular difference in their ability to support larger numbers of people at higher standards of living, in happier settings, and in more stable societies.

KA: If I'm not pressing you again, you said you wanted to distinguish between education and training.

NH: That fits here. Training consists basically of the repetition of procedures in doing a task. We prepare somebody to do that by imparting and perfecting skills. The ability to do fundamental writing, the ability to do simple arithmetic, the ability to use language, the ability to use computers, the ability to use tools and instruments, and the ability to use one's hands are all skills. Many societies have become relatively proficient at training in these and similar areas. Early man did a lot of training in chipping flint or grinding stones. It was a practiced skill. We still have lots of skills acquired by repetitive practice and training.

Education, on the other hand, is of a different order. It is related to the fact that the only constant thing in the universe is change. We might say we can *train* people for the present, but we *educate* people for the future. I am not saying that skills we train individuals to use cannot or will not be used in the future. It is simply that if a person uses basic skills in ways he has been trained to use them—with no awareness of the currents of change—that person can very easily be left behind. The ability to *adapt* skills of performance to changing circumstances requires something that transcends pure training, and that transcendent element is education.

What is so appealing and attractive about "training" over "education" is that we can recognize almost immediately many of the results of the specific training because we are dealing with observable and measurable skills, manipulations, and performance. To say, however, that we should educate efficiently is probably a contradiction of terms. Education cannot be replicated in identical, repetitive, measurable forms because education, unlike much training, is a completely individualized process. And it can be a real puzzlement because its value is often not immediately or even ultimately observable. Its value is more inherent in the individual; it creates *potentiality* in the person. Individuals learn. They can be induced to learn, but not forced to understand. The measurement of the efficiency of education is so difficult in part because different individuals are differently endowed and motivated in how they learn. Different individuals simply do not come to understand in a uniform or predictable way. They certainly don't do it on a fixed time schedule.

While the imparting of skills can be observed, judged, and measured, the degree of learning, the level of understanding, the mental processes, and the applications of understanding are so individualized that they cannot be readily observed or measured. The value of learning or understanding particular information or subject matter content is affected by time, circumstances, maturity, and the feedback a person receives about what is meaningful, usable or rewarding at different times in one's life. Consequently, learning and understanding, the basic components of education, are not efficient and neat processes. They are definitely not quantifiable.

Remember the old saw of the teacher saying, "Neatness counts"? That may be true in training, such as in penmanship, but not necessarily in imparting thinking skills.

KA: You mention the teacher. . . .

NH: Yes. It is this distinction between training and education that makes teachers so important in the educational process. Yet, even while they are so highly valuable, they are not the most valuable component of education. What is most important is the individual student. Teachers can try to induce students to learn, they can nag students to learn, they can challenge students to learn, but in the end they cannot *make* a student learn one damned thing. And without learning, regardless of how good the teaching is, education does not take place.

KA: You used a current platitude a while ago: "The only constant in the world is change."

NH: Call it a platitude if you want to, but it's true. The only constant thing in the universe *is* change. And education is preparation for dealing with our increasing awareness of that. Everything is changing. The universe is expanding constantly. Stars are formed and die. New material is formed, old material is used up, our planet is under constant change. And our perception of what we know is changing faster than anything else.

If we were talking, say, a hundred years earlier, we would not have known enough to say this. We now know that the solid earth shifts. We know that tectonic plates are sliding against and over each other, some moving upward to form mountains and others downward into the interior of the earth. We know how water evaporates and condenses. We know that the oceans have deep currents in them. We know that ocean temperatures in the southern hemisphere influence weather in the northern hemisphere. We know that physical change occurs constantly. Biologically, every living cell forms, works and dies, and every dead cell converts to carbon dioxide and water. We know that forms of life evolve and mutate.

We know as well that societies change. No society or nation is a constant. Ethics change, morals change, customs change, laws change, so society is definitely not constant. As R. H. Tawney has said, even capitalism was anathema for centuries and was slow in getting established. The Church argued that competition had been designed by the devil as a substitute for honesty. And the charging of rent for the use of money—or interest as we call it now—was unchristian and sinful. Nothing on this earth or in the universe is constant.

So why is that important? Because the body of knowledge a student learns in high school doesn't persist longer than a few years. Much of it is changed drastically with time. The important point about change is that a true education, above all else, prepares individuals for accepting and adjusting to change, in fact, to search out change. We need to prepare people to understand how to use their heads to confront the things that may be true in the future. In order to do this, education simply cannot focus just on what is known today.

Yet we cannot teach what we do not yet know. That's certainly a given. And since what we may find to be most useful will be something we will only come to understand at some point in the future, education has to consist in large part of preparing students to expect change, to be able to adjust to change, and to be able to think for themselves within circumstances and a new base of knowledge which cannot be known today.

KA: All right, so now you've got an educated society, or a society with a fair number of people prepared to accept change or deal with change in their world. That alone is not going to bring about technological advancement.

NH: Maybe not, but it's a precondition for discovery and technological change. Remember where I started. We train people for today; we educate people for the future. A society has to be receptive to change in order for science and technology to have a breeding ground.

KA: Well, that is certainly true if we look at the Middle Ages. For almost a thousand years in Europe, no one dared to be different or suggest a new way of doing anything at risk of society putting him to death.

NH: That's the point. And even after change is underway you still find Luddites trying to wreck the innovations.

You want modern examples? Even with the imperfections of the GATT treaty and NAFTA, free trade is an economic plus for all nations.

Now within a technologically advancing society we need a core of scientists and engineers. However, one of the most serious problems in the United States today in the field of science—though not so much specifically in engineering—is the declining pool of bright and motivated students. The most motivated science students still select scientific fields of study, but there are many others who are just as bright who are choosing to go into more service-related fields such as medicine, veterinary medicine, law, accounting, or business. And I'm not saying that's bad; it's just that they're not coming directly into science or advanced levels of engineering.

One reason students are making this choice is that there is a great tendency, particularly among younger people, to want to get a quick return on what they do. They are not accustomed to waiting for their rewards. They don't tolerate delayed gratification very well. Some students are driven primarily by money, but it seems all of them want their service to be quickly noticed and recognized. We could say their curiosity about nature loses out to their impatience. The difficulty of becoming a scientist is that by and large a scientist does not see the fruits of his work recognized for many years or even many decades, if ever.

The second reason is that science is receiving unjustified and unfair criticism as a consequence, ironically, of our having used science and technology so effectively to reduce the risks of being alive. It is too often assumed that society should be risk-free and, perversely, precisely those people who provided for increased longevity are the ones who are being blamed as self-serving scientists and engineers who produce the accidents that cause injuries to people. That idea of a risk-free society and the popularized view of the "pointy-headed scientists" who presumably produce industrial accidents are enough, unfortunately, to discourage the interest of some students from entering careers in science.

KA: You refer to the disdain for "pointy-headed scientists," but I'm not sure we should put all the blame on the general public's inability to appreciate the work of our scientists.

NH: Oh, I completely agree with that and I want to talk later about how we need to educate our non-scientists about what we do and how we contribute to society.

But you are certainly right. Part of the lack of understanding by politicians and the general public is our own fault. Scientists engrossed in their research and speaking their own lingo rarely think about whether somebody else might need to understand what they're doing or talking about: quarks and red-shifts, the Doppler effect or neutrinos or megabytes or genotypes and phenotypes or tectonics. These are names and shorthand terms used in fields of science, and every field is full of them. Some are ancient terms and some were coined last year. The Latin and Greek labels on Vesalius's anatomy charts are still used by the medical profession and by physiology professors because they've become standardized.

To those in a specialized field of science, these terms serve as a form of shorthand, believe it or not, yet indiscriminate use of them can alienate those who are unfamiliar with them.

And we're also an easy target. When a scientist is looking at some small area of ignorance about nature or the way things behave, his research can be made to sound absurd or ridiculous. Examples used to be publicized each year when Senator William Proxmire made his Golden Fleece Award to the most bizarre and mystifying government-funded research projects his staff could find.

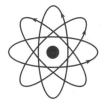

Science and Technology Today

KA: You've spoken frequently about the relationship of science to technology, with the idea that science feeds technology and technology provides new products or processes for the consumer.

NH: Science is nothing more than our knowledge and understanding of the natural universe. At any moment science is both finite and universal. In other words, what science is here in the United States and what is science in Timbuktu or Paducah, is the same. Ignorance is also universal but, in contrast, it seems to me it's also infinite. We don't have to worry about running out of ignorance. There always will be more ignorance than knowledge. Likewise, we do not need to worry about learning "too much."

Science, or knowledge, proceeds by grappling with ignorance of the universe or nature. It is important for us to remember that *people* are also part of the universe. And even though we are an infinitesimal part, we are probably influenced more by human or social elements than by the entire rest of the universe. But we have been egotistical enough to put ourselves above nature, to look at and talk about the rest of the universe as though we are not actually part of it. Yet it is clear we cannot do this. Even if we are godlike in our awareness of the universe and in the miracle of our existence among the inanimate mass of the stars, we are nevertheless an integral, if unpredictable and changing, part of the universe in which we exist. As such, we need to understand ourselves if we are going to try to predict consequences of different actions or behaviors.

The fundamental point I am trying to make is that science is a *creation* of man since all knowledge has to pass through a human mind. That means our view of science is man-made and very much controlled by others around us. That is, our knowledge of nature and the universe is very much affected by society, as Galileo—or Servetus, or even Semmelweiss—quickly discovered.

KA: Let me stop a minute to remind myself. Galileo looked at the moon and stars with a new telescope and documented that Copernicus was right; the sun and the universe did not revolve around the earth. Galileo had to recant his findings to save his life. And even at that he was excommunicated. Servetus did early anatomical work and discovered that the blood passes through the lungs. And Semmelweiss believed that medical students going with unwashed hands from the corpses in the dissection room to examine pregnant women spread childbed fever and death.

NH: And the Vatican did not reinstate Galileo into the Church until 1989.

KA: And Servetus was burned at the stake by John Calvin, and Semmelweiss was derided as a fool.

You make the point very well: science is clearly a creation of man—and is controlled by man. New knowledge of how the world works can't be of any use if society is not willing to accept it. Obviously, technology is a creation of man as well. Just where does technology fit in with science?

NH: Most people are not very precise about the difference between science and technology. We frequently hear the phrase "science *and* technology" as though the two formed a unit. But there is a very distinct and important difference between them. They have different procedures, different concepts and pursuits, and are based on different ideas about the roles and functions for mankind.

As I have explained, science is the inquiry into ignorance. Technology is an *application* of science—our understanding of nature.

Because it can be applied anywhere in the universe that humans can go or see or even speculate upon, science is universal; but scientists and engineers are not.

They are very localized. They belong, as we believe chauvinistically, to *us*, depending on who "us" happens to be, whether Russia, Germany, Japan, or the U.S. It is important to remember that it is the scientists and the engineers who are the most important part of the technology underlying a region's economic vitality, not the *science*–or *knowledge*–as such. This is because science without scientists and engineers is a total abstraction. The knowledge of science and the processes of technology exist only as ongoing activities in the minds of scientists and engineers, who convert science to technology and apply the technology to human use.

KA: I can't help but be reminded of the observation made by Boorstin in *The Discoverers* about why Geneva and London became centers for clockmaking around the middle of the sixteenth century. It turns out that these were the two locations in Europe where Protestants were not being persecuted. Men could come there to practice their trades related to making clocks and watches with a minimum of religious or political interference. And as a critical mass of tradesmen and early scientists began to accumulate, they began to attract others.

The word soon got around Europe that if you knew anything about metallurgy, or making small gears or escapement devices, or running a metal lathe, or making small springs or blowing glass, there were good jobs in those two locations. Early science, early technologists, and early investors got together to create new consumable products for use in their home countries and for export. And out of these processes some people started to make money and others continued to advance the frontiers of knowledge.

NH: A modern lesson from that is not to offend or persecute scientists and scholars if you want to have a technologically advanced society. Without good scientists and engineers, a country will not have functional and usable science. Without adequate numbers of scientists and engineers, there will be very little conversion of science to technology—and even less conversion of technology to its many uses—that is, into humanly beneficial and economically rewarding purposes.

KA: That invites the obvious question: Where does a country get its scientists and engineers?

NH: In the United States, universities are the sole source of scientists and engineers, the absolutely sole source. Except for the few we import, we get them from no other place except the universities. That wasn't always true, but it is now.

That alone should tell our political leaders how essential it is to provide support for higher education in order to generate and promote economic development and vitality. Universities provide the adequately educated scientists and engineers without which a society does not have the slightest chance—short of accidentally running across a diamond mine or gold mine or another thirty trillion barrels of oil—of remaining in the economic race. This means we have to take our human resources—our people—educate them in our schools, bring them into our colleges and universities, educate them again and further, and in the process adequately expose them to scientific and technological research.

KA: But you were going to talk about the relationship of science to technology.

NH: All right. Let's start with the role of technology. Technological innovation consists of a fresh idea, plus a development of that idea, plus a translation of the developed idea into technology, plus a transfer of the technology to human use through a lot of other steps, including financing, manufacturing, marketing, sales, distribution, exporting, and now disposition and cleanup. In other words, all the lines of action by which a fresh new idea ends up as a product to be consumed by the public are at the far right-hand side of the process of translating scientific knowledge of nature into technology usable by humans. On the far left side are all the ideas, which originate in discoveries about the laws of nature.

Despite any beliefs to the contrary, inventions do not occur without the use of science—in other words, the knowledge of nature. An individual must have background and knowledge before designing and making an invention. Even in jumping to a conclusion or seeing a new connection, the scientist or the technologist using a scientific idea is connecting realistic and at least reasonably workable possibilities, based on knowledge and experience. Mr. McCormick did not produce his harvester and reaper without some physicist or engineer first determining how you take a force that moves in one direction and convert it to a force that moves in another direction. McCormick *used* bevel gears, but he did not dream them up. Somebody had to do that work first.

KA: Let's go back to the fundamental requirement for technological innovation. What is necessary in order for such an invention or innovation to occur?

NH: Innovation depends entirely upon an idea. But an idea is not a *sufficient* condition for innovation although it is a *necessary* condition. This holds for all fields, whether technological or artistic or literary. No bright idea *alone* has caused a breakthrough in art, music, literature, or technology. But no breakthrough in art, music, literature, or technology has occurred *without* a bright new idea. New sonnets, operas, musical instruments, architectural designs, advancements in health care, energy use, and recycling, would never become available without such ideas. By itself, an idea is not a usable innovation. It has to be taken up by a person who can do something to it or with it.

Technology is stretched to its limits by *use* on the right in the process I described, but it is fueled by *science* from the left. That is a simple model of the innovation process: science, technology, and use, reading from left to right, with numerous loops and feedbacks, constitute the system. People on the utilization side may say, for example, "I wish I knew what metal I could use in order to do thus-and-so." That works its way back to a technologist who then searches the scientific base of knowledge about nature and develops a new or appropriate alloy or polymer or whatever for use in the process of innovation.

When we look at science on the left—as preceding technology in the middle and at a great distance from utilization on the right—we see that science spins and spins, imparting momentum to technology only fractionally on the edges, in the same way a fluid clutch does. Science rotates and feeds upon and into itself all the time, and there is often no discernible connection between science and technology, or any conceivable or foreordained use for a scientific theory being developed or proven.

Scientists are fascinated first and foremost with pushing back the blanket of ignorance, rather than with making suggestions about how their findings might be put to some practical use. Scientists who do stop frequently in their investigations to consider practical applications have little time left over to discover new scientific information. They usually cease being good basic scientists, even if they do become good entrepreneurs or applied scientists.

On the other hand, in contrast to the fluid-drive connection between science and technology, there is a solid connector between technology and use. This one is focused and firm. From the point of view of the utilization of knowledge, the "adapters" usually know precisely what they want, or what they need, and the technologists know what can be done. We might say that these two relate to each other through a set of solid gears in contrast to the fluid drive connecting the body of knowledge and the technologists to the left.

KA: Go ahead. I'm staying out of your way.

NH: Let me give you this same model in even simpler form: "S" for science, on the left, is connected to "T" for technology, in the center, by a broken line. The "T" is then connected to the right with "U," for uses, by a solid line ($S \dashrightarrow T \rightarrow U$). The broken line indicates a more tenuous or at least unpredictable relationship. The solid line says there is a clear and understood connection from both directions, from both technology and those converting science and technology into uses for mankind.

In the area of science, on the left of my model, the scientist does not know what he wants specifically. In a sense, a scientist follows his nose. Enough scientists following this process can contribute to increased understanding of the nature of the universe, increased growth in the body of knowledge. From that point, some technologist or engineer may see a particular use for a piece of this knowledge and will convert it to technology.

KA: I remember an example that L. Joe Berry, one of your colleagues from microbiology, used to talk about. Sometime before the turn of the last century, Sir Frederick Hopkins took a special interest in finding out what chemicals made butterfly wings so brilliant in color. Can you think of anything legislators or Congressmen could have more fun ridiculing or poking fun at than a professor spending his time pondering the colors of butterflies? Such research would surely be a candidate for Senator Proxmire's Golden Fleece award today. Anyway, Sir Frederick in England and some researchers in Munich spent years studying the chemistry of pigments in butterfly wings, little knowing what purpose their research would eventually serve. It turned out that the pigments Sir Frederick was studying were related to folic acids, and the structures of those folic acids, defined chemically, served as the base for research on vitamins and metabolism.

NH: That's a perfect case of a true scientist going about his research unfettered by the need for practical applications, and later having his discoveries about nature prove useful. Roger Williams, who did much of his work on vitamins at the University of Texas, was also a scientist who followed his nose. His colleagues kept telling him he should be studying rats and animals if he wanted to make any progress on the study of metabolism. But he spent his time on yeasts. And out of that work he broke open the secrets of metabolism in animals and humans, work which subsequently led into research on genetics as well.

KA: But the point Joe Berry made was that there is often a lapse of time between doing research and finding its practical applications. The span of time was over sixty years in the case of relating pigments of butterfly wings to research on vitamins.

NH: And an understanding of exactly this kind of flow from science to technology to human use is critical in policy decisions relating to the application of public money to the support of science. Providing financial support only to research projects which solve specific problems is gambling. In the first place, anyone with that expectation is going to be disappointed. Second, the use of public money for this purpose is going to divert some good and capable people away from doing what they do best, following their noses, and into doing what they do not do well—that is, trying to solve specifically designated problems.

KA: So you're saying that science can't be organized?

NH: Not in the sense of directing scientific scholars in their basic research and areas of interest, or in the sense of anticipating desired scientific outcomes, because science is a constantly rotating system where people scramble around in their ignorance, produce some information, some understanding of nature. Later, somebody else is likely to say, "That is exactly what I have been looking for." Just like the Joe Berry story about butterflies and vitamins.

The body of scientific knowledge, that is, an increased understanding of the nature of the universe, is produced by research. In the United States, this process takes place mainly at universities, although it is also true that some valuable research is conducted in industry and in government laboratories. Such directed research is undertaken because of *perceived* ignorance, in contrast to the basic scientific research taking place in universities, which is more motivated by *pure* ignorance.

KA: I presume that's more than a play on words. Can you develop that idea a little more?

NH: The difference is that research in industry and government laboratories usually is focused on a specific problem, which is why I use the term "perceived" or "organized" ignorance, because it is specifically identified and often has an expected or defined outcome. Pure ignorance, which motivates science in the universities, is driven in all directions by the researchers who do not presume to know what we do not know. Part of the purpose of basic research is to identify *what* we do not know.

KA: Now you've made another distinction, that between the basic or pure research scientist and the applied research scientist.

NH: Both kinds of research—basic or pure versus focused or applied—are important. However, let me distinguish between them.

Many research scientists wonder why systems act the way they do, so they go to work to try to find out. Let's leave those scientists for the moment busy at that. Elsewhere we find engineers looking for a process that will produce a ceramic, but they are unable to do this because there is no known way of determining the conversion of material A to material B. At that point, the engineers may consult one of the scientists studying a "why problem," a scientist searching the base of knowledge to see what is known about why it is difficult to convert material A to material B. Then the scientist may come up with a gadget or a system which says that this is how you might turn A into B, and this is what the conditions have to be, etc. Or he may produce something that will substitute for material B that serves just as well. Then the technologist can try to design a method for doing it. In a way, the applied scientist has a foot in both the science and the technology camps. Sometimes he's more in one than the other. The engineers I've referred to in this situation are also close to being an applied scientist.

Next in the utilization phase, other people will determine whether the process is economically feasible, whether it has a serious impact on the environment, whether there is sufficient demand to justify the cost of production, whether venture capital can be obtained to try it out, and so forth.

KA: Government can sometimes be indirect in trying to get what it wants. An agency may fund a research project and then, in evaluating its progress, may criticize the researcher for not having accomplished his original intent, or it might hint at not continuing his funding, or actually cut off project support, in attempting to direct where the agency wants the researcher to be going.

NH: The only evaluation on a basic research project that should ever be made is whether the researcher performed high quality scientific work—and, of course, whether he or she spent the money properly. When a researcher is following his nose, he may start with one hypothesis and end up doing his most meaningful research on something somewhat different because the new area looked more promising to him along the way than his original plan. To hold a researcher to a narrow focus of investigation, even when it is originally self-defined by the investigator, is too constraining to permit researchers to do their best science. The most deadly kind of research evaluation would be to measure whether every scientist receiving a grant actually kept focused on the original project proposal.

The vast pool of ignorance we are grappling with is not amenable to being organized into neat pieces for exploration. Our ignorance at the edge of what we do know is not specifically definable or describable. We often tend to stumble over pieces of what we need to know. Sometimes it is important for us to explore *now* what we can only *later* come to understand.

Plans may help. Hypotheses certainly do. That's why we start with research proposals. But it is doing good science, not just keeping your nose to the grindstone or wearing blinders, that makes the difference in turning up new knowledge about how nature works.

KA: A little earlier you distinguished between technology and science, saying that science is done primarily in the universities. Where is technology carried out?

NH: While most basic science is done in universities—the National Science Foundation states that over fifty percent is done there—technology is predominantly done in industrial and government laboratories. And, of course, nearly all of the conversion of technology to its useful applications, with the exception perhaps of health applications and some policy studies or economic analyses of the consequences of the innovation, is done outside of universities.

KA: Do you want to talk about the allocation of manpower and resources in your model?

NH: That's important. We can look conceptually at how science, technology, and utilization relate to one another.

Science, on the left of the model, requires staffing, basic support facilities and specialized facilities, and, inevitably, money. Technology requires facilities, a larger number of staff in addition to the technically inclined, and money. The utilization phase at the right requires even more staff and more facilities, and much more money because of the many steps involved, including demonstration, design, fabrication, manufacturing, marketing, distribution, and the sales force. It also should include costs related to environmental preservation or cleanup and restoration. And let's not overlook the costs of recordkeeping and arguing and settling disputes and law suits.

The utilization end on the right is very expensive, technology in the center is less expensive, and science, on the left side, is even less expensive, perhaps ten percent or less of the total.

But if the utilization process draws upon technology and technology must draw upon science, then this gives us some perspective on the need society has to maintain adequate funding for science. There is an inescapable flow from left to right, from science to human uses. With science so clearly one indispensable part of this process of innovation, expenditures on science should obviously be looked upon as an investment.

It is impossible to tell at the time of actual investment in scientific ventures exactly what the money will produce directly—or even indirectly—that is of use to our citizens, most of whom want direct results, if not now, then very soon. And the fact is that finding a direct application for a specific research project may take twenty or thirty years or more. This process calls for more patience than most politicians have, especially when spending public money.

The only way we can show the relationship between a given kind of science and a given human utilization is to look *backwards* from right to left to discover how a particular innovation was dependent upon a certain scientific base of knowledge. This means that looking upon financial support for science as an investment *qua* investment, that is, to make money right now, is a pointless exercise. However, those who have faith enough to put money into scientific research on the premise that it has in the past produced things of use to mankind, will be rewarded. What will work is to support the existing system, for it produces the people who will produce the solutions—*in due time.*

KA: You implied earlier that some societies might go about this differently.

NH: True. Some countries use a different model for investing in research. Rather than using public money, they encourage industry to invest in support of basic science. Or they use public money in government-supported laboratories to finance basic research. The problem with those arrangements is that seldom can the people directing and controlling the funds keep from directing the basic scientists toward what seems to them the most profitable use of their time. And there goes the game. The value of basic research is lost. This is why "hands-off" public support of basic science in the universities has worked so well in the U.S. It is also why this country houses so many Nobel laureates.

KA: You said it is difficult to predict applications of scientific discoveries or research when the work is originally being done, but that by hindsight we can see how specific products tie back to technological transfers of scientific discoveries. Can you give an example?

NH: The National Science Foundation did a study about twenty years ago to do precisely that. TRACES was the acronym for the study called Technology in Retrospect and Critical Events in Science, and it took a backwards view of five new items in use at that time. They set out to find the connections over the years between scientific ideas and discoveries and their application in technology and manufacturing of the products in use. One of those was magnetic recording tape and recorders, and another was the oral contraceptive pill. The ideas and findings drawn on to end up with recording tape were traced back almost a hundred years, through a whole network of small and totally independent theories and discoveries. These dealt with materials, magnetic theory, electronics, and frequency modulation, all of which contributed to and were essential to developing and producing something we all take for granted now as essential in our lives—audio and video tape.

The oral contraceptive was traced backwards from about 1960 to the middle of the last century. The starting point for the 1960's "mission-oriented" research to develop a birth control pill was the basic research done earlier in the physiology of reproduction, hormone research, and steroid chemistry. I don't remember exactly, but it seems to me the original research had to do with fertility studies, and that new knowledge has today given us the ability to prevent conception.

In both cases, mere bits and pieces of the enormous accumulation of scientific knowledge, in several different fields, were plucked off the revolving wheel of knowledge in my model and used as needed and available. The end use had not been envisioned or imagined by those early scientists doing the ineluctably essential, but not at the time clearly applicable, early studies. They were the ones I mentioned earlier who wanted to find out why systems work the way they do.

KA: They had followed their noses.

NH: Right. But not everybody understands the need for patience in making the connections between new understanding about nature and its later application. Nor do they always see the connections between scientific knowledge and its application, even with the benefit of hindsight.

There was another government report actually called Project Hindsight, published by the Department of Defense about 1965. Their conclusion was roughly this, "We have looked backwards and we don't find any weapon currently in the field that was based on research done by the Office of Naval Research, the Army Research Office, or the Air Force Office of Scientific Research." And all of those three research offices were less than twenty years old.

Clearly the report was critical of investments in scientific research, because they could find no connections. Had they gone back farther they would have said the same things found by the TRACES study.

They just took too short a period to note the connections to previous science. Radar, especially important in World War Two, came out of an understanding of microwave reflection. Sonar, the proximity fuse, the nuclear bomb, all came out of scientific research, not out of weapons research.

Even if the two studies seemed to come to opposite conclusions, I think they both lead us to the same lesson: "*You have to be patient.*"

KA: And you're also pointing out that the 1990s are not the first years in which government studies and government policy makers have said, "We have to be more pragmatic and directive about the research the government funds."

NH: That's right. The key point is that in an earlier day, when a piece of research was underway, we could not have traced the original pieces of science *forward*, to the right on my model. Once something is produced it seems in *retrospect* to be the inevitable step toward utilization of those separate segments of knowledge about how the universe works. And similarly today we can't trace forward successfully or accurately from some specific scientific work being done to predict some discrete future idea or innovation that will end up being useful to mankind. Such prescience is just not in the nature of the system.

KA: How do we get those appropriating money to understand this?

NH: Ah, therein lies the basic problem for justifying public support for basic research. In looking at the turning wheel of science we cannot relate accumulated pieces of scientific knowledge forward into a technology that is certain to eventuate in a use or a product.

Our inability to do this raises a basic concern among political leaders: "What is happening to the taxpayers' money I am voting to use for this purpose?" This is predictable because politicians prefer money they spend to show solid returns before their next election. And most run for reelection every two years, an infinitesimally short period in the process of discovery, conversion to use, and manufacturing of a product and the creation of jobs.

KA: Someone said we have twenty-year problems with five-year plans and one-year appropriations. You've added the two-year election cycle.

NH: Well, that is how the scientific and technological system works. If it seems somewhat slipshod and stochastic, it is. It cannot be changed without wrecking it. But if science is not supported, technology will not have the new basic knowledge of nature to draw upon. And technology has no other source but science for new ideas which can result in innovations or new uses to improve the lot of mankind.

KA: Your linear model has been questioned. Does innovation really occur this way?

NH: We have to consider the connections from the massive storage of all scientific knowledge to convergence in the very thin line leading through technological development and production toward a single end result. Furthermore, there are a great number of different individuals participating along the way. The connections between the knowledge and its application and the connections of the different people involved only *appears* to be linear by looking back in time.

Moreover, how truly linear is any process that draws from a scientific base of knowledge those elements that may have been known but unused for decades? Or how linear is a process that starts and stops and stumbles through application and development and commercialization over many years? This model accurately describes the paradigm by which the vast and growing knowledge mankind has about nature finds its way into products or processes useful to mankind.

KA: You talked about the relative staffing requirements of each component of your model. How well off are we in being able to staff the different functions?

NH: Well, I have to say that, other than some deterioration of facilities and equipment and a prevalent (though not universal) belief that we do not attract enough bright, motivated young Americans to the study of science, science in the U.S. is in quite good shape today. Probably ninety percent of all scientists who ever lived are alive today, and probably ninety percent of all scientific knowledge has been accumulated in the past thirty or forty years.

Science is moving very fast these days. Because of the availability of exquisitely sensitive instruments of measurement, we are able to do experiments today we could not have done a very short time ago. It is getting to be true that nothing is too fast or too slow, too big or too small, or too hot or too cold to be measured. We have very inquisitive, very sensitive, very reliable instruments. We also have more powerful mathematics and faster ways to process the mathematical computations.

Because we can do new kinds of experiments, theorists who had less to work with over the past few decades have now been apprised of whole new sets of information. And they have developed new theories, which have produced new experiments, which have produced new theories, which have produced new experiments and new instruments and new mathematics, which we must hope will continue on ad infinitum.

KA: So even though young people are not choosing scientific careers as frequently as you would like, science is in pretty good shape?

NH: Comparing today to 1955, we are certainly better off. Comparing 1995 to 1975, we're not as well off. Compared to 1935, we are infinitely better off.

KA: How about the criticism of overspecialization?

NH: Fields of science, which began perhaps after Newton's time to become compartmentalized, now are coming back together since we are getting to know nature so much better. In the more recent past, an individual could not be just a zoologist, but he had to become a vertebrate zoologist or even an upright vertebrate zoologist as contrasted to a four-footed vertebrate zoologist. Or an individual might be an organic chemist or a physical chemist, or a micro (as opposed to an organismic) biologist, but these compartments are falling apart. With more knowledge and understanding of nature, scientists see the edges that defined breaks between fields of study or specialization becoming fuzzy, and they are returning to the "natural philosopher" idea of Newton's era—and with a lot more knowledge.

In sum, the scientific community is in good shape. What we have to watch out for is that we don't founder suddenly due to some hiatus in adequate funding. Recovery of momentum after a gap in support would be slow, and we'd spend a long time just getting back to where we are now.

KA: Why do you say that?

NH: Our brightest people would find something else to do, and it would be hard to lure them back. The threat of another down time in their future careers would scare many of them off permanently.

KA: You said earlier that the scientists and engineers are more important to a society than is the science itself. Do you want to elaborate?

NH: The engineers and some scientists are the translators. Most scientists serve as torch bearers who continue to keep the flame of knowledge and the pursuit of new knowledge alive. The engineers are the means by which the materials and theories and knowledge of science can be converted to technology and utilized by mankind. They need each other and each needs a foot in the other's camp.

The existence of the knowledge and the repositories of the knowledge will serve no useful purpose unless portions of that knowledge are passing through the minds of actively involved scientists and engineers. Ideas do not originate in pools of accumulated knowledge; they originate in the minds of individuals as they awaken to new connections and relationships. Knowledge of nature as science and as an abstraction which exists only on the written page will return to the cosmos upon the biodegradation of that page. Only that science processed through people's minds can serve mankind.

Even accidents among scientists and technologists that can produce beneficial effects cannot be recognized as potentially beneficial without a prepared and experienced observer. It is the scientist who understands what he or she is witnessing and can translate and interpret what is observed.

KA: How about the financiers?

NH: If the ranks of scientists and engineers can be joined by entrepreneurs who are willing to take risks and put up venture capital, then new ideas will find their ways into creating economic vitality. That's the way it happens now. If the ranks of engineers and scientists did not exist, entrepreneurs alone could not produce economic viability. We would all then see economic development occur in other regions or nations where good scientists and engineers are keeping science alive and healthy.

As I said earlier, without new scientific discoveries and advances in technology, we cannot produce enough natural, agricultural, or mineral resources to generate the economic surplus needed for a high standard of living for rapidly increasing populations. We have no choice but to enhance our human resources through education. We need to prepare and nurture our cadre of scientists and engineers whose minds are prepared to make the connections required both to expand knowledge and to translate knowledge into technology and into beneficial uses.

KA: There was an MIT critique of American business five or six year ago that said we put too much effort on basic research and "break-through" research. They said more emphasis ought to be placed on turning good ideas into products, on downstream technical development.

NH: We've never just been in the discovery business in our universities. Even our engineers don't just invent things. There is a natural coalition among different faculties that leads into manufacturing and producing things. And there is a mutual dependency of the financiers and engineers and scientists and business people. They are always looking to each other for ideas or money. We just need to find ways to nurture that interdependency.

The bottom line for public policy is this: To reap the fruits of the agriculture of technology you have to have the seeds of science. And the producers of the seeds are the scientists. And the sowers of the seeds are the engineers and technologists and applied scientists. We need to educate and support the seed producers as well as the seed sowers.

# Technology Transfer

KA: We hear so much about technology transfer as a way to speed up the uses of science. How does that work?

NH: It's sort of a simplistic expectation—as though it is something you can push into happening by marketing scientific information to engineers and manufacturers and users. Technology is spoken of and conceived of sometimes as an organized activity that simply needs to be put in place so it will begin functioning. People speak of technology transfer as a process for advertising what science has available to be developed by technology and put into manufacturing processes for the market and for mankind's use.

KA: But technology transfer does take place.

NH: Of course, but not often as an organized strategy. Wherever the transfer of scientific knowledge and technology to specific manufacturing or consumer use does occur, it is usually far downstream from where the ideas originated in scientific investigation. Technology transfer takes place when those ideas have worked their way through what is technologically feasible in order to come to a point of manufacturing some usable product or developing some usable service. In actuality, so-called technology transfer occurs all along the way from the origin of scientific ideas to the development of manufacturing processes. There is no simple, identifiable point in the process where ideas can be picked out and promoted into a manufacturing process to create usable products. Panning for ideas must occur along the entire stream between science and manufacturing, for there is no single point where it is more profitable to mine for such ideas than others.

KA: But we cannot deny the existence of an expectation that we can somehow *force* the translation of knowledge into some application or use.

NH: Expectations even from the highest levels are not automatically realizable or even reasonable. I think a more practical concept of technology transfer would be the recognition that there is indeed a possibility of connections occurring between *developing*–as contrasted to *developed*–ideas. Connections can occur among developing ideas further upstream from the manufacturer or user, as well as above technology in the process and closer to the origin of scientific ideas.

Let me give you an example. In many manufacturing processes, machinery or molds have been used to make mechanical parts. In the design stage, especially, prototypes of three-dimensional pieces are needed. A major problem has been how to create inside a solid mold an empty space which is exactly the shape of the piece needed, or how to create the object *without* the mold—in other words, how do we create the prototype by using machinery, when the piece has never been seen or made before?

One solution has been to create two halves of a mold, carve or shape the prototype into the two sides, and then combine the two parts to provide the space in which the casting is made. Another has been to shape and reshape and refine and fine-tune a piece until it finally comes to take the shape of the part that is needed.

A graduate student named Carl Deckard at the University of Texas envisioned combining a computer and a laser to create prototype pieces. The computer is fed a 3-D image of the object needed. Successive thin layers of powdered polymers or other material are laid down, and the laser beam sinters that portion of each layer that is to become part of the desired object, discarding the remaining unsintered powder. That is, by heat it sort of evaporates or welds designated sections of each thin layer of powder to the part being created. The sintered layers, as controlled by the computer, progressively accumulate to leave an intact, welded prototype.

It's an ingenious solution that became possible as Deckard connected different technical capabilities in his mind and brought them to fruition. The process was patented, licensing rights were obtained and it was finally brought into commercial production.

KA: You make it sound like a big part of the process is being at the right spot at the right time.

NH: That's only part of it. It's the right *kind* of person being at the right spot at the right time.

KA: How's that?

NH: Those who attempt to serve as technology transfer experts or "brokers" of ideas usually look for opportunities downstream from scientific discovery. The actual transfer of ideas along the stream of knowledge takes place as part of a creative act. Brokerage in this sense has to be much looser. Involved as it is in making unusual and even unexpected connections among many combinations of vaguely defined word images, brokering does not call much for people outside the science or technology community.

A non-science person cannot play a highly useful role because he or she is not equipped to make the connections among the many vaguely defined ideas and concepts that lead further downstream to manufacturing users. Such a broker or "idea entrepreneur" is usually looking upstream from the user or manufacturing end, and hence much of what he sees is largely incomprehensible or uninterpretable.

This is why scientists not only originate new findings of knowledge; they also contain the minds uniquely prepared by scientific work to recognize possible connections that cross their sense of awareness and catch their attention. The engineer and the technologist are similarly or perhaps even better prepared and conditioned to do that. A technology broker or promoter of technology transfer cannot be an entrepreneurial dilettante; the individual has to be a trained or, better yet, a *practicing* scientist, engineer, or technologist.

Many ideas that move from science through technology into production and then to a consumer or manufacturing use happen almost as if by accident. No broker is needed. A connection is recognized by a properly equipped mind.

KA: Wasn't it Pasteur who said, "Chance favors the mind prepared"?

NH: Precisely. When scientists see something unexpected in a process and recognize some possible practical application, it is because they are prepared to recognize the meaning and significance of the unexpected event in terms of science or technology.

It is an example of someone picking some piece of information out of the vast storehouse of knowledge about nature that's been revealed by past scientists. This occurs only if the person witnessing the event is equipped, experienced, and prepared to recognize what he or she is observing and possibly knowing where to find some other piece of knowledge which may be needed. Or being able to recognize a relationship that some observer is uniquely qualified to see.

Sir Alexander Fleming, who saw what penicillin mold was doing in some spoiled petri dishes, is the classic case of the mind prepared to recognize something of significance in a chance or unusual event. Not only was he a well educated and practicing scientist, but he had also had previous experience with a spoiled culture dish. Once, by accident, he had let some of his nasal discharge drip into a culture dish, and later he observed that bacteria were not growing near his "contamination." From that he had gone on to identify the enzyme in human nasal mucous that retards bacterial growth. So you can see Fleming was uniquely positioned and uniquely qualified to discover penicillin—or more precisely, to recognize the effect of the contamination of a particular mold in some petri dishes. Specifically he recognized that staphylococci were not growing around the edges of the penicillin mold. You can see that the accidental contamination he witnessed and interpreted as a useful occurrence in nature would have been lost on all but an infinitesimally small number of scientists, much less on the population as a whole.

KA: You said much knowledge is still discovered by accident. It seems that being *inside* the scientific community is essential to recognizing the meaning of such accidents when they occur.

NH: Absolutely. And inside the scientific community *continuously* and as an active participant. Sometimes the connections in observed accidents are not necessarily between pieces of scientific information. Sometimes it is a matter of seeing an application or a marketing opportunity. ScotchGard fabric protector originated from an accident. I have been told that a researcher at 3M corporation spilled a beaker of chemicals on the clothing of some trainees she was working with. Several days later she noticed that the spots on their garments where the chemicals had spilled were cleaner than elsewhere. We'd have to say she stumbled onto this discovery through her sense of observation, seeing both the results of her "accident" and then seeing a marketable use for it.

Another such case occurred at Raytheon. A technician noticed that some radar equipment he was experimenting with had melted several chocolate bars lying nearby. That was the origin of the microwave oven.

These and similar cases are not accidents so much as they are cases of serendipity. In the loose fabric of scientific knowledge, discovery is not conducted in a planned manner to identify holes that need to be patched, or, to keep the metaphor going, to be woven shut. Some holes get filled in, but not because scientists as a group function with a "central mind" that directs or plans how and when discoveries should be ordered. The relationship of all the threads to one another are not formal, are not precise, and are never assured. The revelation of knowledge is more evolutionary than programmed. To use my fancy word again, it's to a large extent stochastic.

But I don't mean to give the impression that the field of science contains the only discoveries of any consequence. Artists, as well as scientists, create works that remain unappreciated during their lifetimes. Van Gogh has much company for doing long, arduous, creative work with no timely recognition.

KA: So it would be accurate to observe that just as great poems or musical compositions are rarely written by the untrained artist, great scientific discoveries are not likely to be revealed to someone untrained in the sciences.

NH: Nor is it likely with a scientist who is not working continuously at his field. When some aberration, quirk, accident or serendipitous event takes place in the presence of an untrained observer, it is like the proverbial tree falling in the forest. Air and sound waves may reverberate, but if there is no ear to hear the tree, the sound goes unnoted. Not only is the technological application of scientific knowledge among the inhabitants of this planet an anthropogenic process, but also it is only possible for the relatively few, prepared anthropoids to do it.

Let me give you an example from the field of medical research. There is something called MLD, metachromatic leukodystrophy, which is a fatal disease of small children. Out of all the contributions to identifying and trying to figure out how to treat this disease, the identification of its trademark in patients came about because a young physician had spent an unusual amount of time doing urine studies as a medical student and again as an intern. His unique preparation was that he had spent a lot more time looking at urine sediments under a microscope than had most interns or researchers.

When a pair of twins with this serious nerve disorder came to his attention, he theorized that he should be able to obtain and stain a sediment from their urine that would be a marker of the disease. Indeed, he did find the marker he was seeking, but it turned out to be an entirely different sediment and was of a different color than what he was expecting. Had he not been uniquely qualified and experienced in studying urine sediments, he would not have formulated his hypothesis. But maybe more significantly, had he not been so uniquely prepared, he would not have spotted that other, differently colored sediment and recognized what it meant when his original expectations weren't met. The continuous and current exchange of information is critical. That young medical researcher relied on articles, conversations, consultations with colleagues, and new, unexplainable surprises in his studies and specimens to make this unique discovery.

Scientists must constantly, in their minds, juggle masses of vaguely defined word images, concepts and ideas to see what connections are possibly worth investigating. They need also to know when some "hole" in the loose fabric of knowledge about nature—or their special piece of it—gets filled in, and they do this through constantly "playing around" in their field of science. Playing around is what keeps them current—and excited and motivated, I might add, about their profession, their life's work.

KA: You've been using the metaphors of a stream carrying knowledge through development and then to users, and that of a fabric of knowledge. Could we go back to your metaphor of the gears and the flywheel to describe the place of technology transfer in that model? Could you talk some about the technology transfer process in that context?

NH: It is important to note that the possibility of the transfer of *science to technology* and then to its later use is far easier to recognize than it is to see and bring to fruition a connection between *technology and use*. That's why there are a heck of a lot more patents than ever end up being produced or used. Think about the poets or writers you referred to earlier. How many ideas for novels do you think went through Trollope's or Hardy's minds compared to the books they actually did complete and publish?

As the flywheel of science continues to spin around, technologists and engineers and other applied scientists see ideas and possibilities vaguely taking shape on the flywheel. From time to time they pick something off as it takes form and they try to imagine how it might be made useful.

A couple of examples may help in understanding this. When scientists Otto Hahn and Friedrich Strassman split the atom in the 1930s, many scientists, engineers, and technologists immediately saw the potential uses of this new discovery to the production of massive amounts of energy. But the actual use and application of this information was still a few years away. Nuclear *fusion* is another example. The possibility of fusing hydrogen atoms to release enormous amounts of energy was recognized, but in all the resulting research, the billionth of a second in which the fusion process has been sustainable has resulted only in uncontrollable nuclear explosions. Other than for war, no use of nuclear fusion to mankind has yet been manageable. The possibility of harnessing this energy, while theoretically feasible, still lies years away in actual use—if it is ever achievable.

The creation of the nuclear bomb was certainly one of history's fastest transformations of science and technology into use, all because the need to create a weapon was so great. A lot of American and European scientists believed that German scientists were trying to produce a similar weapon for use by the Nazis. We can only recoil at the thought that all the world would have sunk into a new dark age had the first nuclear bomb been dropped on London or sent by V-2 rocket in 1944, or even early 1945.

And the fact is that a number of scientists in Chicago were moving their families out into the country in 1941 because they felt certain that Hitler was going to get the nuclear bomb first, and that he would soon use it on us. There was good reason to believe that. Germany had Werner Heisenberg, one of the world's foremost nuclear physicists, working on their bomb. The American physicists knew they were just arriving at where Heisenberg had been three year earlier, so they assumed they were far behind where the Germans were in making the bomb.

The Nazis even talked publicly and on the radio about making such a bomb. We had evidence at that time that in Norway, Germany was increasing production of heavy water, which can be used in the making of a nuclear bomb. And Czechoslovakia, under German rule, had Europe's only uranium mines. There were even American scientists and European refugee scientists here making macabre bets among themselves on when the Germans would get their first bomb.

KA: You said that not everything discovered is necessarily useful. That ought to be obvious to any thinking person, but maybe it would help to understand that better if we had an example of something that seemed to have potential but has never produced a usable product.

NH: Or has not turned out to be of use as of *now*. But I can give you an example that fits what you're asking for. About forty years ago there was a great deal of interest in something called "iron whiskers," small accumulations of iron in a very dense and consequently very strong form.

As elements come together, that is, as atoms align themselves, there are typically vacancies. That is, as nature forms, the lattice of atomic structure has some holes in it on the atomic level. The "iron whiskers" had far fewer holes or vacancies in their atomic lattice, so that this kind of iron had properties different from ordinary iron. That is why it was of such interest. The hope was that we could find a way to produce these "iron whiskers" prolifically. But it just never panned out. To my knowledge it was never economically producible and consequently is not commonly worked on today.

KA: So in sum, you don't see much place for the person specializing in technology transfer.

NH: Not as such, not as an outsider to the process. There indeed may be a place for brokers along the stream between science and technology and commercialization or use. But such a person is going to have to be a combination of scientist, engineer, technologist, and entrepreneur. He might go further if he concentrated on any one of those fields.

KA: But if you've ever met some of those brokers who have been successful on the interface between science and technology or with the investment bankers, you know that they are eager workers. They're looking hard for the next strike.

NH: Yes, they're like a cat that's just eaten a small bird and is busy looking for another one.

But with such a demanding background of education and experience needed to be successful, the person also will have to be somewhat specialized. For example, there already are such brokers at work in the health fields, but even there we would not likely see the same broker working in even the closely related fields of pharmaceutical drugs or clinical instrumentation. There is too much to know in order to be prepared to see the vaguely defined word images across widely disparate fields. Yet when possibilities are identified, such brokers can play a meaningful role if they can explain the possible product, can give assurances about the sound underpinnings of the scientific knowledge upon which the product is to be based, and can talk to engineers, designers, and financiers. Such a facilitator can be invaluable to all parties downstream, from the scientists at the headwaters to the salesman moving the finished product at the user's end of the river of progress. But as I said earlier, it is ironic that those who contribute the original ideas often will be among the least interested in what the broker is promoting.

KA: Or they may be long dead.

NH: That's certainly true. Perhaps the reason for the greater difficulty in making the meaningful transfer in the model I have set up is the solid gear driving technology to application and use. The fundamental questions that make the gears mesh between technology and use include the following: Can the technology effectively be applied to use? What is the cost? Can sources be found to pay for the cost? Is there a consumer of the product if these other questions can be resolved? Is there a substitute product that is cheaper to provide? Will the clean-up costs make the undertaking impracticable?

KA: As you said earlier, the costs at this end are much higher, and far more people are required, than at the other end.

NH: Financing plays the major role at the stage of commercialization because of the additional question concerning the consumer. This ties back to another question: Is there a sufficiently consistent user over a long enough period of time to get the financing back out of the investment and make a profit? This is the question which will determine whether the venture capital will be made available. If it is deemed doubtful, all the potentiality, feasibility, and practicability of some new product will go for naught. Many a good, patented and licensed idea never goes anywhere for this reason.

Again, the transfer of the scientific idea of splitting the atom is a good example. It went through all these processes. The incredible costs initially required were provided by government in a time of crisis when it was thought that there was a race between the United States and Nazi Germany in applying atom splitting to weapons. Since then, the application of nuclear energy to peaceful uses has been found to be effective, and sources of finance have been found despite the high costs. Consumers willing to support the return on the capital investment in nuclear plants materialized. The additional question of whether a consistent user over a long enough period of time is available has been raised with the subsequent concerns about the environment and safety. Three Mile Island and Chernobyl are quick to come to mind.

There are also the clean-up costs when we're dealing with toxic isotopes with half-lives of thousands of years. The shut-downs of newly operational nuclear plants on Long Island and in Washington State are examples of unforeseen "costs" to technology already placed in use.

KA: So what does this tell us about public policies related to technology transfer?

NH: The lesson from these principles is that we need to be careful not to over-organize science simply because we are so anxious to find quick applications and uses. It is my view we need to keep the small bits and pieces of information and knowledge coming in large quantities from individual scientists. We need to promote the education and preparation of young scientists as an integral part of all funded research projects. This should be our primary goal: strengthen the pool of scientists and engineers. The rest is by-product and will naturally follow behind the major goal.

We need to remember that science is produced best in a random and unorganized, or at least loosely organized, system of the pursuit of knowledge. Consequently, we need a hands-off approach in government funding of scientific research. The important overriding principle is that we should never apply the same cost or efficiency measures to science that we apply to the highly organized process of manufacturing technology at the production end of the technology transfer process.

 Science and the Non-scientist

KA: You have often talked about the need to teach science to non-science majors . . .

NH: Or to people who don't think they are interested in science, or will ever use it in their everyday lives.

Science, or the exploration of nature, is another art form, one purpose of which is to improve human comfort and longevity. The other art forms improve culture, increase self-satisfaction, and help us to understand ourselves and live together.

And science that is done for its *own sake* is wonderful for the scientist. It's fun, it's rewarding, it's fulfilling. Science done for its *effect* is for the greater populace. That's where it has its impact. The beauty of this is that they are both the same science.

The time has come when more people need to understand what scientists do and how science affects us all. Non-scientists are the vast majority of the voters. Scientists cannot continue to neglect their responsibility to teach non-scientists just how crucial science is in all our lives.

KA: In what way have non-scientists been neglected?

NH: We scientists have been so involved in doing science—that is, in extending understanding and knowledge and in finding ways to prepare our successors—that we have lost sight of those who are paying for all this. We have forged ahead, assuming that everyone else understands and fully endorses all the good things we have been doing.

The bulk of the population pays for the few of us to practice science. Some of those who are paying for it see us having fun doing it without understanding what we do or how it really affects them.

KA: And I'm sure you can sense the public's suspicion toward scientists. Some few actually feel scientists do more harm than good.

NH: That's the point. The problems we have trying to explain and maintain an adequate level of support for science, combined with public distrust about scientific methods, I believe, are in part consequences of our not paying more attention to science education for non-science majors.

KA: "Trust us; you need us," certainly doesn't sell very well with the voters anymore.

NH: Or with their representatives in Congress or in state legislatures. Those of us who work in scientific fields know what we've accomplished and what our discoveries have meant to improving the human condition, but *knowing* our own contributions is no longer enough.

The mind-set of too many scientist is that you cannot properly understand science if you don't understand higher mathematics. I don't buy that. You may not be able to *do* science without competence in higher mathematics, but you can certainly *understand* it and even *enjoy* it, just as you can appreciate art or music without having to be an artist or a musician.

KA: Then how can we go about teaching more science to non-science majors?

NH: We have to do it on several levels. The current approach is to water down the content of a course for those who can't follow higher mathematics. We should not give up on explaining a phenomenon because the only way we can conceive to explain it is through math. That is not a good idea.

Those of us in science with all the access to the storehouse of knowledge about nature ought to be able to discover ways to elucidate what nature is about without our explanations being dependent on higher math. Concepts, models, relationships ought to be generally explainable and understandable without having to exclusively use formulas and numbers. New technologies offer a great opportunity. If computers and video screens can now make it possible for us to see things that don't exist, surely we ought to be able to use computers to help people visualize objects and processes that are too small to see or happen too fast or too obscurely to observe.

KA: Then why haven't we done it?

NH: I'm guessing now, but I suspect it's because we scientists haven't seen it as an important task. We'd rather spend our energies on things that are more fun.

KA: A friend of mine said recently that in his state everything is politics except politics, and *that* is personal. It certainly seems as if everything has a political component, and that makes it more important for people to understand how the political process works, how our institutions of self governance work.

NH: You're reading my mind. With more and more of our everyday lives being touched by science and technology, we need to be sure that we all understand what science does and how it affects us. And with science being politicized, the need for people to better understand us becomes evident.

KA: So we're back to how to do it.

NH: In higher education, I think we need a required course at grade 13 or the first year of college. Such a concept should convey the point we discussed so often earlier: that science is simply an understanding of nature.

The course should begin with a brief introduction to formal logic, then build around the four components of nature: materials, forces, time and space. I would try from there to show some underlying relationships among those components, then to develop the idea of how and why and when science began to split into its various disciplines. I'd make clear that, generally speaking, those who want to delve deeply into science must hunker down in one of those narrow trenches.

But if you're not going to be a participant in scientific work, all you need is a general understanding and some sense of relationships and connections.

KA: Anything else?

NH: Yes, I'd cover what we've been talking about—how science relates to technology and society, as this relationship seems to be generally not understood, or at least underappreciated.

I'm not saying that non-scientists *can't* understand the relationships. No, the criticism rests with *us*, the science community, for not figuring out how to communicate what we do and how it affects how we all live.

KA: This is what you would do at the higher education level. How about in earlier schooling?

NH: I would wager if we did a better job in our colleges we would create the capacity for teachers to fairly and competently present science in grades K through 12. I believe many elementary and secondary teachers can't make a fair presentation of science today because they are fearful of science and math.

And doing a better job at the college level would help parents and families and kindergarten teachers not to be afraid to discuss or explain science to children.

The greatest quality that young people bring with them to school is their curiosity about everything around them. Too many times we fail to incorporate this quality into children's learning patterns—or we even see it as threatening to classroom order, so we try to ignore it or stamp it out.

What I would *not* do is this: I would not begin discriminating among fields of science before the fifth or sixth grades. I would treat nature as a unit. I would give explanations of nature not based on complex ideas or theories of physics, chemistry, geology, or astronomy. Explanations should be as holistic as possible. Teachers should not hide behind statements like, "This is a law of physics or of magnetic fields," but rather should use statements like "These things attract," or "These things repel."

The point is, knowledge about nature should be kept together as a unit until the students, of their own accord, begin to see that to learn more about a particular aspect of nature you have to separate and specialize. But this is not so peculiar to science. We also do it with grammar; we don't start out teaching students theories of grammar. And we do the same in mathematics. You don't start with esoteric or abstract ideas or theories, but with concrete, elementary problems.

KA: You have mentioned to me before the kinds of questions youngsters ask out of their curiosity and how they should be answered. For example: "Why is the sky blue when it appears from outer space to be black?" or "Why does water freeze from the top down rather than from the bottom up?" or "Do hot things absorb cold or do cold things absorb heat?" And the most persistent question of all, "Why," which follows on so many explanations we give to children.

NH: I'm not a grade school teacher, but I'd say children's questions should be answered, as closely as possible, in *their* language not the scientist's language. We should be able to work with grade-school teachers to fit scientific explanations into the students' own language and perception levels, to nurture their curiosity and keep it alive.

The persistent "Why" question from children is merely the most palpable evidence of their overwhelming curiosity. Everyone can be comfortable giving a set speech if they rehearse it enough, but teachers of curious children can't rely on set speeches. They don't have the privilege of rehearsals. They have to be able to roll with the punches and constantly respond to the unexpected. That's why we need to make teachers feel more comfortable in their ability to handle the kinds of questions you just put to me.

Furthermore, it is not just that teachers should be able to explain something or answer questions; they also need to know how to motivate the students to learn, on their own and in groups, about how nature works. They need to guide students into doing things that cause them to learn the answers for themselves. Unless teachers feel comfortable in their ability to deal with any kind of pitch that comes at them, they will revert to pat answers or hide behind abstract and unintelligible explanations. And that's exactly what turns off students' curiosity.

We also need to do a better job of explaining science and technology to our grade-school children so that, as citizens in the future, they will have some idea of what's happening around them, what's affecting them, and what's being done to make their lives better and where these improvements come from. I believe everyone should have some sense of where science fits into the affairs of the human race.

KA: Because the children will someday be voters who will decide on levels of support for science in the future.

NH: Don't make it sound so self-serving, so "science-serving." There are many other reasons to provide today's children with a firm scientific foundation. Citizens are going to have to decide for themselves, based on what they read and hear, whether global warming really is taking place. Or whether the ozone layer is being damaged. Or whether the oceans are being over-fished. Or whether to worry about the rain forests. Or acid rain. Or the risks of nuclear power. Or air pollution. Or animal experimentation. Or genetic engineering.

We can prepare people to make informed decisions on more and more complex issues, or we can let them form their opinions by listening to radio and TV talk shows. If people cannot recognize hogwash when they hear it or read it, they may think it's pork roast.

The last time I counted there were more non-scientists than scientists. And since in politics we decide issues by counting people's opinions and votes, we scientists had better help the non-scientists understand the importance of science to their lives and futures if we really do believe what we do makes a difference to the quality of life.

KA: We are back to where we were a few minutes ago, on the political decisions being made about support for science. For all the dangers that may come from poor political policy-making, science has fared rather well so far from federal policies and support.

NH: By all means. I don't mean to take anything away from the enlightened positions legislators and various administrations have taken in recent decades. I am one of the principal beneficiaries of that system of support.

As I have said, I have lived through the golden age of science during the history of mankind. That golden age was to a large extent defined by the policies of the U. S. government in support of basic and applied sciences. But if we are going to ensure that the government continues on its enlightened path, we need to be certain that the people sending representatives to our state and federal governments have a better idea of what science is and why it deserves their continued support.

KA: How did we get to the point where the federal government plays such a major role in furthering science in this country?

NH: To answer that, we need to look first to the European antecedents. The charter of the Royal Society of London, founded in 1660, states that its purpose is to improve natural knowledge. To me that means understanding nature—or "science," as I have defined it earlier.

Then in 1744 Benjamin Franklin founded the American Philosophical Society for the purpose of promoting useful knowledge. Do you see the difference? Here, in a nutshell, is an example of the pragmatism of the new American colonies versus the old British or European view.

Whether actually based on that philosophy of utility or not, the policies of the federal government from our very beginnings until 1952, with the creation of the National Science Foundation, focused on providing support for specific (and usually pragmatic) scientific needs. An analysis of the scientific projects which were funded before 1952, indicates that the U. S. government invested very little money in research in basic science up to that date. Some of the programs that were funded by Washington as a means of focusing on the scientific needs of the country were the Public Health Service (founded 1798), the Coastal Survey (1807), the Smithsonian Institution (1836). The National Institute was founded in 1840 but it failed two years later. And then, in the midst of the Civil War, the Morrill Act was passed in 1862 to create agricultural and mechanical colleges for the practical application of science and technology. In 1863 Lincoln signed the charter creating the National Academy of Sciences. The new Academy was to advise the president and the government on how to handle certain aspects of the war such as munitions, weapons, clothing, food, and health.

After the Civil War, the government continued to promote focused efforts in science. In 1879 the Geological Survey was created and in 1891 the Weather Bureau. By the time World War I began, the National Academy of Sciences had become mostly an honorary society, but it created the National Research Council in 1916, again to give advice on how to apply science and technology to a war. The Cancer Institute, the first of the institutes of health, was created in 1937. With our entry into World War II, the government again needed help from the science community, so the National Defense Research Council was created. This agency contributed to developing the nuclear bomb, radar, sonar, the proximity fuse, rockets and other war-time applications of science.

What I'm trying to do is illustrate the propensity of the U. S. government, before 1952, to fund extremely focused organizations, some of which were very short-lived. But in 1945, Vannevar Bush published the report of a government committee he had chaired. In *Science, the Endless Frontier*, he made a case for the need for support of basic, in addition to applied, research, in order to produce future scientists and engineers as well as to establish an ever-increasing pool of basic scientific knowledge. He suggested establishing a "non-mission agency." The National Science Foundation was created in 1952 as the result of the recognized need for this kind of agency, but the fact is that NSF does have a mission: To secure the pursuit of basic science, to strengthen educational institutions as the places for carrying out that kind of research, and to enlist and prepare more scientists and engineers.

So we might say that there is practically no evidence that the federal government undertook any programs to support science for its own sake from around 1781 to 1952. But since the creation of the National Science Foundation, the government has pursued aggressively the enlightened goal of securing properly trained *people* to make the scientific *system* work. There are even those who have come to consider securing an educated people as the *objective*; and the resulting science and its applications and uses are merely the *bonus*.

KA: But even within the government-funded agencies there was some investment in securing the people needed from the universities.

NH: That's true. But with the possible exception of the Office of Naval Research, the agencies ordinarily were not interested in basic scientists. The Office of Naval Research, however, unlike other federal agencies, did become interested in people for the long haul, not just to help with their immediate problems.

KA: So for the last four decades the non-scientists in the government, although largely unversed in science, have been doing a good job in supporting science.

NH: True. For the last forty years this country has pursued through the federal government a more enlightened objective in science than ever before in our country's history. But in the current political mood of the country, there is a tendency to want to get back to practical applications, to set narrow objectives, to direct scientific research to specific goals.

Although this is not altogether bad, since it still indicates support for science in some form, it is nevertheless a worrisome trend. The idea of abandoning the goal of preparing our successor scientists is misguided, particularly when measured against the immense advancements in science and engineering we have had in the last forty years. Such a short-sighted approach will inevitably diminish the role of our universities as the place where scientists are formed and where they do their most creative and productive work.

This is why I say, in all seriousness, that we need to get started teaching both our children and the non-scientist adults about the values and uses of science.

# Big Science and Little Science

KA: Do you want to talk about "big science" versus "little science"?

NH: There's a place for both. Science, as knowledge about nature, doesn't exist to bring about agglomerations of scientists to do what has come to be known as "big science."

But let me start with "little science." Most new knowledge has accumulated through "little science," that is, small projects pursued by individuals or a small team of individuals. Little science provides multiple outlets for inquisitiveness. And it gives us pretty quick answers to specific questions.

Is it worth the money being spent on it? I'd say yes. Others might ask, "Since so much scientific knowledge lies waiting for application and technological use, why keep adding to the scientific pool of knowledge as fast as we are?"

I would reply that the expressly usable results or outcomes of science are merely the *dividends* of funding research projects. The greatest long-term—as well as the immediate—benefit from funding research is to increase the quality and quantity of our pool of scientists, mathematicians and engineers.

Look at it this way; the research process produces several things. It ignites the interest and commitment of students as they see professional scientists grapple with challenging problems. It maintains and stimulates creativity in the scientists. It increases the store of knowledge about how nature works. And in addition to all this, research provides the ultimate bonus of economic gain and a more comfortable standard of living for more people.

As I mentioned earlier, knowledge must be processed through minds that are prepared and equipped to do that kind of work. So by definition, the greater the number of scientists and engineers congregated in a particular location, the more knowledge about nature there will be churning around there. Those places with a large number of scientists are most likely to produce new ideas and new applications of knowledge, resulting in by-products such as patents, copyrights, licensing rights, new businesses, or improved new products and new jobs.

KA: But with big science you have aggregations of scientists as well.

NH: True, but there's a difference. In big science, we bring together a large number of scientists to work together on *one* problem. But with little science you can have lots of scientists working on their own areas of interest without any formal coordination. The scientists and engineers working in that kind of setting cross-pollinate each others' ideas, so to speak. Take Einstein's theory of relativity, for example. It didn't come out of big science, it came out of his ability to theorize. His was one of the very small number of major new discoveries that come to us each century. And it came out of little science. If Einstein hadn't discovered the theory of relativity would someone else have done it? Sure.

Big science is intended to deal with questions you have no chance of answering with little science. Big science usually involves big instruments or arrays of instruments spread around the world. The human genome project is one example—a four-billion-dollar or more project.

Do we want to see the edge of the universe? You need a Hubble telescope. Do we find out whether "Lucy" is or is not anthropogenetic? How are we related to the burst of life and new creatures of the Cambrian period? Do we want to know what matter was like in the universe fifteen billion years ago, in the first few seconds of the life of the universe? High energy colliders cost billions of dollars, and they are run by a single team of scientists for this pretty narrow purpose. This is big science at its best.

What difference does it make whether we ever answer these questions? There are a lot of people who feel it is indeed important to press ahead. Some feel that being interested in answers to questions like these is what makes us different from all the rest of the dust in the universe. But as we have seen from the termination of funding for the super conducting super collider, there are others who feel we should not press ahead on this particular project.

KA: When you began your studies in science, there wasn't a lot of big science underway.

NH: Not much. There were General Electric Laboratories and the Bell Telephone Laboratory and a few others like them. There were astronomy observatories, oceanographic and atmospheric research centers and other systemic research labs. But I didn't know about them at the time.

KA: How did you get interested in science?

NH: I took elementary chemistry in high school, and I was intrigued by it. As a junior, I took analytic geometry and calculus. As a senior, I took elementary physics. And I got a lot more depth than high school kids do today in those subjects.

When I was young, the prevailing wisdom was that if you studied science beyond a certain level you became a medical doctor. If you were wildly curious about nature and how it worked, you were expected to satisfy that need in some other way *after* you became a doctor.

But when it came time for me to go to college, I made my decision based on personal experience and interest. People asked me back home, "Why are you doing this rather than becoming a doctor?" I didn't have an answer except to say I was interested. When they asked me, "What are you going to *do* with it," I didn't have an answer. If they had pursued with me what they were really thinking, they would have come right out and asked me, "Why are you wasting your parents' money this way?"

I was lucky. My parents let me follow my nose. And I came into the study of science just when a few people were beginning to recognize that scientists could be valuable even if they did not pursue medical studies. My career has spanned the golden years of science in this century.

KA: And to think you could have been an M.D.! Was there anything specific that guided you from general science into chemistry?

NH: It might sound strange to kids today, but what got me interested was a development in surface chemistry in the middle 1930s. For the first time, it became possible to count molecules by measuring the area of a film on the surface of water. A man named Irving Langmuir and a woman named Katharine Blodgett took a tray of clean water and put a solution of sodium palmitate, a soap, and acetone on top of the water. The acetone evaporated, and the sodium palmitate molecules, being only slightly soluble in water, floated around on top, upright, and connected to each other laterally in a very thin layer only a molecule thick. By measuring the cross-sectional area of the film and knowing the cross-sectional area of the palmitate chain, simple arithmetic could tell us how many molecules were in the film.

The problem we'd had with counting molecules up to this point in time was that the number of molecules in 18 grams of water is 6 times 10 to the 23rd power. You can't count that high. But with this discovery you could take a small *surface*, and you could count off a thousand or two thousand molecules by an area because they were of a single thickness and because we knew the cross-section of each molecule in its orientation on the surface was 20.5 angstroms squared.

During the great drought of the 1930s, the U.S. needed something to reduce evaporation of water. This thin cover or film they had created worked fine on small bodies of water—as long as you didn't have sizable waves.

It might sound strange to kids today, but this is what got me interested in surface chemistry. It was the first self-assembly of molecules in laboratory chemistry that was noteworthy and documented and corroborated, although nature has been self-assembling molecules forever.

KA: So it wasn't a fellowship or grant that directed you?

NH: No, but that certainly has become the case today with many young scientists and prospective scientists. Money speaks, and money directs. It has changed personal decisions and motivations. We are inducing students to enter particular areas of study by offering them scholarships and fellowships. A felt shortage in the science or engineering or business community can be affected by money.

I am not sure this is all to the good. Young kids have no shortage of curiosity about nature and how things work. It is later, when they face the need for a livelihood, that they lose their drive to satisfy their curiosity and instead pursue other personal interests or needs. They are asked, as I was, "What are you going to *do*? Your life is ahead of you." Then the availability of money can influence them in ways that might not be as good for them personally—or for science as a whole—as if they exercised their own judgment on where their true interests lie.

KA: But very shortly after graduation you did end up in big science. Your wife, Gene, has told me you could not even tell her what you were doing.

NH: You're referring to the years I spent on the Manhattan Project. Well, in the first place, I did not have a major role developing the nuclear bomb. I was a young teacher at Virginia Polytechnic Institute at the time I got a call to come up to New York. Among my students I had some select soldiers who'd been sent back to college to develop technical capacities the country needed. In that first group I had ten soldiers. We got them going in chemistry, math and physics. Most of them ended up in meteorology, and after the war some went back to school to earn PhDs; in other words they were a great group. But the second group of 100 to 150 did not turn out to be as intellectually curious, to put it mildly. As a matter of fact I think I got a bunch of malcontents and maybe misfits who had been sent to us from their units to get rid of them. They were hard to deal with and I frankly was discouraged. At that point I enlisted in the Navy, but before I could report to Cornell for gunnery school training, I got the call to come to New York City. There I met these two guys, Zola Deutsch and A. C. Loonam. They needed another guy and they'd heard about me somehow. I was maybe thirty.

When I told them I had a Navy appointment, they said forget it. I got busy and never did resign my commission. But anyway, I never heard any more from the Navy. And Deutsch told me I couldn't bring my family; I had to leave them in Blacksburg, Virginia, and I couldn't tell my wife or anybody anything about what I would be doing.

So there I was on the twelfth floor of this building, which, incidentally, was the same floor with General Groves—although I never had anything to do with him. Our job, the three of us, plus a lady named Kelly, was to be the Materials Division of the project to develop one of the processes leading to a nuclear bomb.

KA: And what did you do specifically?

NH: My job was to visit various laboratories where they were conducting research on the materials used to separate uranium isotopes by gaseous diffusion. This was just one of the methods possible that had been proposed to separate U235 from all the other isotopes. It turned out to be an approach which was effective.

One of the materials was a minute mesh made of nickel. The process was to flow uranium hexaflouride, a gas, through tubes made of very fine nickel mesh. Each time the gas flowed past this membrane you got a minutely enhanced separation with lighter molecules to the outside and heavier ones on the inside. The lighter molecules contained uranium 235, the isotope used in the bomb. The heavier molecules contained the heavier isotopes of uranium. This passing of the gas was done a multitude of times in a cascade arrangement until a suitable purity of U235 was reached.

We used very long tubes, one inch in diameter, made of very pure powdered nickel. These tubes were in bundles of 70 to a hundred to a tank and we had thousands of tanks.

KA: Were you involved in the theoretical end of the work?

NH: No, all laboratory. One of the big problems was that nickel in the presence of moisture will corrode and become solid nickel oxide, which could plug up the holes in the membrane and mess up the separation process. And the gas was extremely corrosive. So the laboratories I was working with were trying to solve this problem.

My job was to visit these laboratories to see how they were coming along, without telling any of the separate operations what the others were doing or how they were approaching the task at hand. The idea was to have them deal with the problem independently. There were places like Princeton, Columbia, and industrial laboratories in Cleveland. And I think maybe Toledo.

KA: So was the Manhattan Project really big science?

NH: Not really. It was more a big engineering job. How to do processing on a vast scale. We were translating science into a product for use, in this case a weapon. You might say it was a highly focused demand for use.

Now with the plutonium bomb, that involved more science. They were dealing with newly produced materials and the use of accelerators and cyclotrons. They had to learn a lot about the chemistry of plutonium. They could make guesses about it from its position on the periodic table, but you still had to measure properties, chemical properties. There was a lot more science in that work.

KA: How did you get to the University of Texas?

NH: Near the end of 1944, I went to a meeting of the American Chemical Society at the Hilton Hotel on 32nd and Eighth in New York City. There I met Henry Henze and Roger Williams, both on the chemistry faculty at the University of Texas. They asked me to come to Austin for an interview. I got there on the day President Homer Rainey resigned after a heated political fight over interference by the Board of Regents and the Legislature in his administration of the University. It was hard for me to find administrators who wanted to talk seriously about an interview, but I liked what I heard and saw in the department. Rather than go back to Virginia Polytechnic Institute to my associate professor's position, I decided to take the assistant professorship at UT because of the research possibilities and the strong encouragement for research there in contrast to VPI.

KA: Are you doing anything right now that might illustrate some of the principles you have been discussing?

NH: One thing I'm involved in is the development of a new design for a lead acid battery for electric vehicles. This one holds the potential for fitting in between where we are now with batteries and where the manufacturers and users would like to be in a few years. The concept is there, and at present the job is to find enough financial partners (a) to manufacture such a battery, (b) to distribute it, and (c) to be sure it will, indeed, be reliable enough to displace current batteries for this use. At the same time, we know that we may be looking at a window of economic viability of only ten to twenty years before this improved type of lead battery is perhaps replaced by a new lithium battery. This fact makes it hard to find risk money or venture capital to get such a new lead battery on the market.

KA: Using your model of science flowing to technology and then to uses and what you just said, you are involved in this project on the far right hand, that of manufacturing and production for use by mankind. Can you trace the connections for this particular research project backward in your model to scientific discoveries or theories on the left side of your model?

NH: The idea of a battery was developed between about 1830 and 1860. Earlier, around the 1790s, Luigi Galvani and

Alessandro Volta recognized that they were dealing with a force different from anything they had ever seen before. Galvani happened to lay a severed frog's leg on a piece of copper lying on a piece of iron and saw it move. He knew there had to be some kind of force making that happen.

KA: Another case of the prepared mind seeing something unusual and following up on it.

NH: Exactly. Soon after that Volta stacked plates of zinc and copper and separated them with wet, salt-saturated cloths, and he got a current to flow out of this stack. A number of people were working to improve on this discovery. A Frenchman named Gaston Plante patented the lead acid battery in 1859, but for the rest of the century the idea of a packagable source of energy languished for lack of any general need for it.

As I recall, there were a few users who needed a stable voltage for some small purpose, such as laboratory power sources, but there was no significant demand. In fact, the article on batteries in the 1911 edition of the *Encyclopedia Britannica* explains how they can be used to ring small bells–that's all there is about batteries. At that time, there was practically no use for them.

Then Charles Kettering, in 1912, designed and built the electric starter for automobiles. Half a century had lapsed between the time of discovery and the time of the useful application of a packaged, rechargeable source of energy.

In the project in which I'm currently involved, we are trying to take those ideas from the middle of the last century and make them work better. Much of the advancement is in the *engineering* of the process, not the basic knowledge or theory.

KA: But we still haven't moved all the way to the left of your model, to the theory of why energy flows out of a battery at all.

NH: The process is this. A material is immersed or situated so it will be made to change from one form to another, and in the conversion will give up electrons. The electrons are dumped into a wire and flow to wherever there is a lower concentration of electrons. The flow of electrons in a metal conductor is a current usable as electrical energy.

Our object with this new lead battery is to increase the flow of power out of it. Some of the elements used to make a battery do this more efficiently are lithium, sodium, zinc, nickel, and lead. The first two of these will offer the most power, because they give up electrons easier than lead or nickel or most other materials. So we look to them as theoretically very important in our search for getting a better and more powerful battery.

KA: So what are the obstacles that are delaying a sodium or lithium battery?

NH: To get a workable sodium-sulphur battery you have to have a very high temperature to make the molten sulphur sufficiently fluid. This requires something on the order of 300 degrees centigrade. So you get a very hot battery and obviously there's the possibility of fire with a battery like that. Also if anything goes wrong with the battery casing and the sodium comes into contact with water it explodes. It may not be the kind of battery you put in thousands or millions of vehicles running at high speeds all over the place or that you park with confidence in a garage attached to your house.

A practical lithium battery lies somewhere out there in the future. Lithium is actually more active than sodium, but it's possible to contain it so that it doesn't get exposed too easily. But if it does get exposed, it starts to burn. We don't know how long it will be before we get a cheaper, more efficient and safe lithium battery into mass use, but we do know we can in the near future produce a more efficient lead battery than those in general use now. The question is, "How long is the meantime? Long enough to justify the huge expenditures and investments to produce the new lead battery?"

KA: And if the venture capital isn't there?

NH: Then we will all get by on the present kind of batteries until the new lithium or whatever battery is perfected. Thus the opportunity is lost to reduce the use of resources by increased efficiencies and to make a new profitable product.

KA: This example also illustrates the point you made about manpower investments. The scientific discoveries took relatively few people, but the engineering, production, and delivery end of the process involves many more.

NH: Specifically, let me mention a few. After you have determined that there is a desire, a need, a market, then you have to get the investors to put up the money. You have to obtain suppliers of raw materials. You need people to run the plant and to hire the workers and keep necessary records. When you begin to produce your widgets, you have to store them, and move them, and deliver them. You have to keep working on the engineering to maintain quality and reliability and improve your product and your process for making it. You have to ensure safety in its use. And with batteries, you have to have a system of recycling because of their potential effect on the environment and people.

In addition, all the time you're working and spending vast sums of money, you face the threat of competition making all your investments worthless. Somebody may find a way to deliver the same product cheaper. Or slightly better. Or do a better job of marketing their product than you do yours.

But remember this, all the activity and benefits and new jobs at this production end, *started* with an idea that tied back to a scientific discovery or theory.

The question on our new battery is whether or not our idea is fundable and if it is, will we have enough time to make money out of it before a better battery comes along.

KA: We're nearly at the end. Anything you would like to add or modify that you said earlier?

NH: Maybe only a couple of things. What I've tried to do is get outside of being a self-centered scientist to see what it is we do that's of some value to others. It turns out to be very simple. Maybe not what legislators would like to hear, but not complicated. Science does not exist in the ether of the universe. Science exists in people. So we have to invest in the kind of people who produce science or knowledge about how nature works. Without science and the resulting technology, it is simply impossible to support six billion people, soon to be ten billion, on this planet.

And I don't want to be misunderstood concerning my model of the flywheel and the gears.

Let me illustrate be restating my simple depiction $S \dashrightarrow T \rightarrow U$ that I discussed earlier, science feeding into technology and technology feeding into uses. There are obviously feedback loops and bells and whistles along the way that make this process more complex than my simple model would imply. Science and technology and applied uses are not so separate as my model may imply. Moreover, the time constraints for the two arrows in the simple depiction I mentioned are different. For a new scientific discovery that becomes useful, the broken-line arrow from S to T may often be as short as, say, five years. Whereas the combined time lapse from S to T to U is apt to run to twenty years and beyond. Engineering and science come together and often share common territory. My model is to try to illustrate relationships, not to imply separations that do not exist in reality.

A last example will illustrate another point I want to be sure is understood. Say an engineering process is needed to pull a fiber of glass at, say, thirty feet a minute while you coat that glass fiber with lead. The speed at which the glass fiber can be pulled and coated and not have it continually break may not be as much dependent on science as on engineering capability and ingenuity. Maybe the lead crystallizes too fast or the rate of pulling the glass causes it to crystallize too fast, and it breaks. The point is this: knowledge of glass and lead crystallization rates and how to control them are both scientific and engineering problems.

The other thing I'd add is that the accumulation of knowledge is a spiraling process. More knowledge stimulates the discovery of more knowledge, the *application* of knowledge also stimulates the discovery of more knowledge. It is no historical accident that the Scientific Revolution followed the Enlightenment, or that the Industrial Revolution followed the Scientific Revolution. That's the only possible order in which they could occur. They fed each other. Those periods illustrate my model of the flow of knowledge to development and then to use and the improvement of the welfare of mankind. Furthermore, the insatiable demand of users, consumers, manufacturers and entrepreneurs for more knowledge feeds on science and feeds back *into* science.

KA: In one of our earlier discussions you referred to legislators calling our researchers "pointy-headed scientists." Why is it that legislators seem to have such a hard time understanding the contributions of scientists and scholars?

NH: Not all of them do—especially after they see what science can contribute. Congressman Olin Teague is a good example. He brings to mind my best example of ridiculed science paying off big for society. This occurred in the mid-1950's in Texas. Two different groups of legislators were debating two problems in the state. One was concerned about what to do about professors, particularly at UT and Texas A&M, who were wasting taxpayers' money to study things like the sex lives of insects and other creatures.

And the other group of legislators was studying what the government could do about the devastating screwworm epidemic that was affecting the cattle in the state. The screw worm lays its eggs in any open wound or in the eyes or nostrils of cattle and other animals. The larvae burrow into the flesh and cripple and often kill the animals or drive them insane. Large numbers of valuable animals died or had to be killed. The epidemic was decimating the cattlemen's herds across the state, but especially in south Texas.

Then the university scientists, involved in their presumed prurient interest in the sexual practices of insects, discovered something truly unique about the screw-worm fly—the female mates only once in her lifetime. With that discovery in hand, the scientists then learned that the female was totally indifferent to whether her once-in-a-lifetime male inseminator was fertile or not. So swarms of sterile, irradiated male screw-worm flies were released in the epidemic areas over a period of time. These still virile but sterile male flies competed well with the progressively declining numbers of fertile males, and the female flies kept laying generations of infertile eggs. That plague of the cattlemen was soon eradicated. And I might add that it was done without leaving any toxic insecticide residue in the environment.

Congressman Teague was so excited about this project and the research that he talked about it every time I came to Washington. I never drank so much Wild Turkey again in my life as in those sessions with him talking about the screw-worm fly.

But there is an important point to this. Had the legislature been able to censor or restrict or direct the research of those scientists in the university, this solution would not have been possible.

KA: Or had the legislature or the Congress failed to fund basic research on insects. One last question. Here we've done a whole book on science without a single footnote. What if someone wants to know your sources?

NH: Perhaps they could go to the library. How do I know enough about something to make my interpretations and deductions and guesses? I'm drawing on a lifetime of experience, and work, and reading, and conversations like this. It is what I have heard and seen and experienced.

Index

"accidental" scientific
 discoveries, 60–62
American Philosophical
 Society, 80
animal breeding, 3–4
Berry, L. Joe, 37, 38, 39
Blodgett, Katharine, 89–90
Boorstin, Daniel, 11, 32
breeding, 3
brokers. *See* technology transfer
 experts
Bronze Age, 5, 7–8
Bush, Vannevar, 81
butterfly research, 37–38
Calvin, John, 31
Cancer Institute, 81
ceramics, 8–9
chemical elements, 4–7, 92–93

Chernobyl, 69
Civil War (U. S.), 80–81
clockmaking, 32–33
Coastal Survey, 80
Copernicus, 30
Cortéz, Hernán, 6
Deckard, Carl, 57
Department of Defense, 47
Deutsch, Zola, 92
"directed science." *See* "science
 on demand"
economics, 10–11, 13–19, 38,
 53, 68–69, 90–91
education, 13–27, 33–34, 42, 53,
 73–78, 88–89
Einstein, Albert, 87
engineers, 18, 25, 31, 40–41, 52,
 54

Fleming, Sir Alexander, 60
Franklin, Benjamin, 80
funding for research, 41, 44–45, 48–49, 53, 68–70, 72–73, 79–83, 98–99
Galileo, 30
Galvani, Luigi, 95
GATT treaty, 25
Geological Survey, 81
Golden Fleece Award, 27, 37
Groves, General, 92
Hahn, Otto, 64
Heisenberg, Werner, 65
Hopkins, Sir Frederick, 37–38
human resources, 17–19
innovation process, 35–38
"instructed science." *See* "science on demand"
Iron Age, 8
"iron whiskers," 66
Kettering, Charles, 96
Langmuir, Irving, 89–90
linear thinking, 10
Loonam, A. C., 92
Luddites, 24
McCormick, Cyrus, 34–35
Manhattan Project, 91–92
manufacturing processes, 57–58
Massachusetts Institute of Technology, 53
materials sciences, 4–6
metals, 4–9, 16–17
minerals, 16–17

mining, 4–7, 16–17
MLD (metachromatic leukodystrophy), 62–63
Morrill Act, 80
NAFTA, 25
National Academy of Sciences, 80–81
National Defense Research, 81
National Institute, 80
National Research Council, 81
National Science Foundation, 42, 45, 80, 81, 82
natural resources, 19
nature, 1, 3, 7, 8, 10, 14, 64–65, 90–91
non-scientists, 26, 58, 71–83
nuclear bomb, 15, 47
nuclear energy, 69
nuclear fusion, 64
Office of Naval Research, 82
oral contraceptive pill, 46
"perceived ignorance," 39–40
Plante, Gaston, 96
policymakers, 9–12, 13–15, 33, 38, 47, 73, 74, 79
population, 9, 10, 14–15
Project Hindsight, 47
Proxmire, William, 27, 37
Public Health Service, 80
"pure ignorance," 39–40
radar, 47
Raytheon, 61
renewable resources, 16

research, 8; applied/focused, 40–42; basic/pure, 40–42, 45, 54
Royal Society of London, 80
scholarships, 73–74
science, and religion, 31; beginnings of, 1–9; "big," 85–103; definition of, 3, 29–30; "little," 85–103; overspecialization in, 51–52; relationship to technology and utilization, 29–54, 58–59, 64–66, 100
"science on demand," 9–11, 41–42
scientists, 52, 72
ScotchGard, 61
screw-worm research, 102–103
Semmelweiss, 30
serendipity. *See* accidental scientific discoveries
Servetus, Michael, 30
Smithsonian Institution, 80
Sonar, 47
Stone Age, 8

Strassman, Friedrich, 63
students, 22–26, 90–91
Tawney, R. H., 23
Teague, Olin, 102
technology, 18–19, 25, 29–30, 34–36; transfer of, 55–70
technology transfer experts (brokers), 58, 67–68
3M corporation, 61
Three Mile Island, 69
TRACES (Technology in Retrospect and Critical Events in Science), 45–46, 47
training, 13–27
utilization process, 43–44, 49–50, 59
vitamins, 38
Volta, Alessandro, 96
Weather Bureau, 81
Williams, Roger, 38
world population, 7, 9
World War I, 81
World War II, 15, 47, 65–66, 81
yeast research, 38

Norman Hackerman served as President of Rice University and holds the title of President Emeritus and Distinguished Professor Emeritus of Chemistry there. He spent twenty-five years at the University of Texas at Austin, where he served as President and Vice Chancellor for Academic Affairs, among many other posts. He is now Professor Emeritus of Chemistry at UT Austin. He is a distinguished alumnus from Johns Hopkins University, a member of the National Academy of Sciences, the American Philosophical Society, and a fellow in the American Academy of Arts and Sciences, the American Association for the Advancement of Science and the New York Academy of Sciences. He has served as chairman of the National Science Board, editor of the *Journal of the Electrochemical Society*, and is author or coauthor of more than two-hundred publications.

Kenneth Ashworth is commissioner of higher education for Texas. He has been employed with the U. S. Treasury Department, the Urban Renewal Administration, the San Francisco Redevelopment Agency, and with the U. S. Office of Education in Washington. A former University of Texas System Vice Chancellor and Executive Vice President of the University of Texas at San Antonio, he has served on the Commission on Colleges of the South Association of Colleges and Schools, the Education Commission of the States and the Southern Regional Education Board. He is a member of the Western Interstate Commission on Higher Education Regional Policy Committee on Minorities in Higher Education, and the Southern Regional Education Board Committee on Educational Quality.